Liquid Chromatography for the Analyst

CHROMATOGRAPHIC SCIENCE SERIES

A Series of Monographs

Editor: JACK CAZES
Sanki Laboratories, Inc.
Mount Laurel, New Jersey

1. Dynamics of Chromatography, *J. Calvin Giddings*
2. Gas Chromatographic Analysis of Drugs and Pesticides, *Benjamin J. Gudzinowicz*
3. Principles of Adsorption Chromatography: The Separation of Nonionic Organic Compounds, *Lloyd R. Snyder*
4. Multicomponent Chromatography: Theory of Interference, *Friedrich Helfferich and Gerhard Klein*
5. Quantitative Analysis by Gas Chromatography, *Josef Novák*
6. High-Speed Liquid Chromatography, *Peter M. Rajcsanyi and Elisabeth Rajcsanyi*
7. Fundamentals of Integrated GC-MS (in three parts), *Benjamin J. Gudzinowicz, Michael J. Gudzinowicz, and Horace F. Martin*
8. Liquid Chromatography of Polymers and Related Materials, *Jack Cazes*
9. GLC and HPLC Determination of Therapeutic Agents (in three parts), *Part 1 edited by Kiyoshi Tsuji and Walter Morozowich, Parts 2 and 3 edited by Kiyoshi Tsuji*
10. Biological/Biomedical Applications of Liquid Chromatography, *edited by Gerald L. Hawk*
11. Chromatography in Petroleum Analysis, *edited by Klaus H. Altgelt and T. H. Gouw*
12. Biological/Biomedical Applications of Liquid Chromatography II, *edited by Gerald L. Hawk*
13. Liquid Chromatography of Polymers and Related Materials II, *edited by Jack Cazes and Xavier Delamare*
14. Introduction to Analytical Gas Chromatography: History, Principles, and Practice, *John A. Perry*
15. Applications of Glass Capillary Gas Chromatography, *edited by Walter G. Jennings*
16. Steroid Analysis by HPLC: Recent Applications, *edited by Marie P. Kautsky*
17. Thin-Layer Chromatography: Techniques and Applications, *Bernard Fried and Joseph Sherma*
18. Biological/Biomedical Applications of Liquid Chromatography III, *edited by Gerald L. Hawk*
19. Liquid Chromatography of Polymers and Related Materials III, *edited by Jack Cazes*
20. Biological/Biomedical Applications of Liquid Chromatography, *edited by Gerald L. Hawk*
21. Chromatographic Separation and Extraction with Foamed Plastics and Rubbers, *G. J. Moody and J. D. R. Thomas*

22. Analytical Pyrolysis: A Comprehensive Guide, *William J. Irwin*
23. Liquid Chromatography Detectors, *edited by Thomas M. Vickrey*
24. High-Performance Liquid Chromatography in Forensic Chemistry, *edited by Ira S. Lurie and John D. Wittwer, Jr.*
25. Steric Exclusion Liquid Chromatography of Polymers, *edited by Josef Janča*
26. HPLC Analysis of Biological Compounds: A Laboratory Guide, *William S. Hancock and James T. Sparrow*
27. Affinity Chromatography: Template Chromatography of Nucleic Acids and Proteins, *Herbert Schott*
28. HPLC in Nucleic Acid Research: Methods and Applications, *edited by Phyllis R. Brown*
29. Pyrolysis and GC in Polymer Analysis, *edited by S. A. Liebman and E. J. Levy*
30. Modern Chromatographic Analysis of the Vitamins, *edited by André P. De Leenheer, Willy E. Lambert, and Marcel G. M. De Ruyter*
31. Ion-Pair Chromatography, *edited by Milton T. W. Hearn*
32. Therapeutic Drug Monitoring and Toxicology by Liquid Chromatography, *edited by Steven H. Y. Wong*
33. Affinity Chromatography: Practical and Theoretical Aspects, *Peter Mohr and Klaus Pommerening*
34. Reaction Detection in Liquid Chromatography, *edited by Ira S. Krull*
35. Thin-Layer Chromatography: Techniques and Applications. Second Edition, Revised and Expanded, *Bernard Fried and Joseph Sherma*
36. Quantitative Thin-Layer Chromatography and Its Industrial Applications, *edited by Laszlo R. Treiber*
37. Ion Chromatography, *edited by James G. Tarter*
38. Chromatographic Theory and Basic Principles, *edited by Jan Åke Jönsson*
39. Field-Flow Fractionation: Analysis of Macromolecules and Particles, *Josef Janča*
40. Chromatographic Chiral Separations, *edited by Morris Zief and Laura J. Crane*
41. Quantitative Analysis by Gas Chromatography, Second Edition, Revised and Expanded, *Josef Novák*
42. Flow Perturbation Gas Chromatography, *N. A. Katsanos*
43. Ion-Exchange Chromatography of Proteins, *Shuichi Yamamoto, Kazuhiro Nakanishi, and Ryuichi Matsuno*
44. Countercurrent Chromatography: Theory and Practice, *edited by N. Bhushan Mandava and Yoichiro Ito*
45. Microbore Column Chromatography: A Unified Approach to Chromatography, *edited by Frank J. Yang*
46. Preparative-Scale Chromatography, *edited by Eli Grushka*
47. Packings and Stationary Phases in Chromatographic Techniques, *edited by Klaus K. Unger*
48. Detection-Oriented Derivatization Techniques in Liquid Chromatography, *edited by Henk Lingeman and Willy J. M. Underberg*
49. Chromatographic Analysis of Pharmaceuticals, *edited by John A. Adamovics*
50. Multidimensional Chromatography: Techniques and Applications, *edited by Hernan Cortes*
51. HPLC of Biological Macromolecules: Methods and Applications, *edited by Karen M. Gooding and Fred E. Regnier*

52. Modern Thin-Layer Chromatography, *edited by Nelu Grinberg*
53. Chromatographic Analysis of Alkaloids, *Milan Popl, Jan Fähnrich, and Vlastimil Tatar*
54. HPLC in Clinical Chemistry, *I. N. Papadoyannis*
55. Handbook of Thin-Layer Chromatography, *edited by Joseph Sherma and Bernard Fried*
56. Gas–Liquid–Solid Chromatography, *V. G. Berezkin*
57. Complexation Chromatography, *edited by D. Cagniant*
58. Liquid Chromatography–Mass Spectrometry, *W. M. A. Niessen and Jan van der Greef*
59. Trace Analysis with Microcolumn Liquid Chromatography, *Miloš Krejčí*
60. Modern Chromatographic Analysis of Vitamins: Second Edition, *edited by André P. De Leenheer, Willy E. Lambert, and Hans J. Nelis*
61. Preparative and Production Scale Chromatography, *edited by G. Ganetsos and P. E. Barker*
62. Diode Array Detection in HPLC, *edited by Ludwig Huber and Stephan A. George*
63. Handbook of Affinity Chromatography, *edited by Toni Kline*
64. Capillary Electrophoresis Technology, *edited by Norberto A. Guzman*
65. Lipid Chromatographic Analysis, *edited by Takayuki Shibamoto*
66. Thin-Layer Chromatography: Techniques and Applications, Third Edition, Revised and Expanded, *Bernard Fried and Joseph Sherma*
67. Liquid Chromatography for the Analyst, *Raymond P. W. Scott*

ADDITIONAL VOLUMES IN PREPARATION

Liquid Chromatography for the Analyst

Raymond P. W. Scott

Georgetown University
Washington, D.C.
Birkbeck College, University of London
London, United Kingdom

CRC Press
Taylor & Francis Group
Boca Raton London New York

CRC Press is an imprint of the
Taylor & Francis Group, an **informa** business

CRC Press
Taylor & Francis Group
6000 Broken Sound Parkway NW, Suite 300
Boca Raton, FL 33487-2742

First issued in paperback 2019

ISBN-13: 978-0-8247-9184-1 (hbk)
ISBN-13: 978-0-367-40211-2 (pbk)

Library of Congress Cataloging-in-Publication Data

Scott, Raymond P. W. (Raymond Peter William)
Liquid chromatography for the analyst / Raymond P. W. Scott.
p. cm. -- (Chromatographic science series; v. 67)
Includes bibliographical references and index.
ISBN 0-8247-9184-3 (alk. paper)
1. Liquid chromatography. I. Series: Chromatographic science; v. 67.
QD79.C454S37 1994
543'.0894--dc20 93-43899
CIP

Visit the Taylor & Francis Web site at
http://www.taylorandfrancis.com

and the CRC Press Web site at
http://www.crcpress.com

Preface

The liquid chromatograph no longer resides solely in the laboratory of the chromatographer but is now part of the many instruments available to both the routine and the expert analyst. It has taken its place alongside the gas chromatograph, the IR spectrometer and the many other instruments that serve in the day-to-day tasks of the contemporary analytical laboratory. However, just as the analyst requires a basic understanding of gas chromatography and IR spectrometry in order to apply those techniques to appropriate analytical problems successfully, so a basic understanding of liquid chromatography is necessary to use this technology in the most pertinent manner. This text has been prepared to present a clear perception of the chromatographic process together with the function of its associated instrumentation, the rationale behind the choice of appropriate phase systems and the procedures necessary to obtain accurate qualitative and quantitative analyses.

The book is written in a simple and direct style and will help the chemist with a good basic training in chemistry to acquire a sufficient grasp of its subject to apply the techniques successfully to analytical problems.

I would like to take this opportunity to thank Mr. Tom Beesley of ASTEC Inc. and Dr. Elena Katz of the Perkin Elmer Corporation for providing instrument data and many application chromatograms involving special separations. Thanks are also due to Supelco Inc., Waters Chromatography Inc., Whatman Inc. and Valco Valves Inc., which were all very generous in providing information on their products as well as sample chromatograms of unique separations. Finally, I would like to thank my wife Barbara for her patience and help in editing the many drafts of the manuscript.

Raymond P. W. Scott

Acknowledgments

The author would like to thank the *Journal of Chromatography* for permission to publish Figures 35 and 37 in Chapter 2; Figure 5 in chapter 3; Figure 5 in Chapter 4; Figures 3 and 6 in Chapter 6; Figures 4, 5 and 6 in Chapter 7 and Figures 4, 6A and 6B in Chapter 11; also the *Journal of Liquid Chromatography* for permission to publish Figures 10, 12, 16 and 17 in Chapter 7 and *Analytical Chemistry* for permission to publish Figure 11 in Chapter 3.

Contents

Preface *iii*
Acknowledgments *v*

1. An Introduction to Chromatography **1**

A Short History of LC 2
The Separation Process 4
The Different Forms of Chromatography 7
Chromatography Nomenclature 9
References 13

2. Resolution, Retention and Selectivity **15**

The Plate Theory 17
The Retention Volume of a Solute 21
Factors that Control the Distribution Coefficient of a Solute 23
Molecular Interactions 23
The Thermodynamic Explanation of Retention 29
Factors that Control the Availability of the Stationary Phase 33

Chromatographic Methods of Identification 40
The Capacity Ratio of a Solute 41
The Separation Ratio 42
Column Efficiency 44
References 48

3. Liquid Chromatography Phase Systems 51

The Production of Silica Gel 55
Solute/Solvent Interactions on the Surface of Silica Gel 58
Mono-Layer Adsorption of Solvents on the Surface of Silica Gel 58
Solute Interactions with the Silica Gel Surface (Mobile Phase n-Heptane/Chloroform) 60
Bi-Layer Adsorption of Solvents on the Surface of Silica Gel 63
Solute Interactions with the Silica Gel Surface (Mobile Phase n-Heptane/Ethyl Acetate) 65
Silica Gel as an Exclusion Medium 67
Application Areas for Silica Gel as a Stationary Phase 69
Silica Gel as a Stationary Phase in Elution Chromatography 69
Silica Gel as a Stationary Phase in Exclusion Chromatography 70
Bonded Phases 71
The Different Classes of Bonded Phases 73
Reverse Phases 76
The Interaction of Reverse Phases with Solvents and Solutes 77
Paired Ion Reagents 79
The Use of Bonded Phases in Elution Chromatography 81
Aqueous Solvent Mixtures 82
The Use of Macro-Porous Polymers as Stationary Phases 84
Polymer Stationary Phases in Elution Chromatography 90
General Advice for the Unknown Sample 91
References 91

4. The Liquid Chromatography Column 93

The Rate Theory 94
The Summation of Variances 94
The Maximum Sample Volume 95
The Van Deemter Equation 97
The Significance of the HETP Equation 105
The Reduced Chromatogram 106
Resolution 108
Calculating the Efficiency Required to Achieve a Specific Resolution 109
Peak Asymmetry 111
Column Selection 114
Column Diameter 116
Preparative Columns 117
Volume Overload 118
References 121

5. The Liquid Chromatograph 123

The Basic Liquid Chromatograph 124
The Mobile Phase Supply System 124
The Gradient Programmer 125
The LC Pump 128
The Sample Valve 138
The Column and Column Oven 144
Detectors 149
Data Acquisition and Processing 152
References 155

6. Liquid Chromatography Detectors 157

Detector Specifications 158
Detector Linearity and Response Index (α) 158
Linear Dynamic Range 161
Detector Noise Level 162

Measurement of Detector Noise 163
Detector Sensitivity or the Minimum Detectable Concentration 164
Pressure Sensitivity 164
Flow Sensitivity 165
Temperature Sensitivity 165
The UV Detector 165
The Fixed Wavelength Detector 167
The Multi-Wavelength Detector 169
The Electrical Conductivity Detector 176
The Fluorescence Detector 180
The Refractive Index Detector 184
The Tridet Multifunctional Detector 189
References 193

7. Sample Preparation 195

Sample Preparation Techniques 195
Filtration Techniques 196
Centrifugation Techniques 197
Concentration and Extraction Techniques 198
An Automatic Sample Extraction and Concentration Procedure 205
A Sample Separation Protocol 210
LC Sample Preparation for Solid Materials 212
Materials of Interest Present in High Concentration 213
Materials of Interest Present in Low Concentration 217
The Preparation of Liquid Samples for LC Analysis 221
Materials of Interest Present in High Concentration 221
Materials of Interest Present in Low Concentration 225
The Preparation of Liquid/Solid Samples for LC Analysis 228
Materials of Interest Present in High Concentration 229
Materials Present in Samples at Low Concentration 232
General Comments on Choice of Mobile Phase 235
Derivatization Techniques 237
Pre-Column Derivatization 238
Post-Column Derivatization 245
References 247

8. Qualitative and Quantitative Analysis **251**

Qualitative Analysis 252
The Effect of Temperature on Retention Volume Measurement 260
The Effect of Solvent Composition on Retention Volume Measurement 262
Quantitative Analysis 265
Peak Height Measurements 265
Peak Area Measurements 266
Procedures for Quantitative Analysis 267
Quantitative Analysis Using Reference Standards 267
The Relative Precision of Peak Height and Peak Area Measurements 272
Peak De-Convolution 273
Reporting Analytical Results 277
References 279

9. LC Applications **281**

Separations Based on Exclusion Chromatography 282
Exclusion Chromatography Employing Silica Gel 283
Exclusion Chromatography Employing Micro-Reticulated Cross-Linked Polystyrene Gels 286
Chiral Separations 290
Interactive LC Systems 296
Dispersive Interaction Chromatography 297
Polar Interaction Chromatography 304
Ionic Interaction Chromatography 309
Mixed Interaction Chromatography 314
Summary 319

Index *321*

1
An Introduction to Chromatography

Chromatography is probably the most powerful and versatile analytical technique available to the modern chemist. Its power arises from its capacity to determine quantitatively many individual components present in a mixture in one, single analytical procedure. Its versatility comes from its capacity to handle a very wide variety of samples, that may be gaseous, liquid or solid in nature. In addition, the sample can range in complexity from a single substance to a multi-component mixture containing widely differing chemical species. Another aspect of the versatility of the technique is that the analysis can be carried out, at one extreme, on a very costly and complex instrument, and at the other, on a simple, inexpensive thin layer plate.

The unique character of the chromatographic method that makes it so useful arises from the dual nature of its function. Chromatography is, in fact, a combined *separating* and *measuring* system. The sample is first separated into its individual components and then, by the use of a sensor with a quantitative response, the *quantity* of each component present can be measured.

Chromatography was invented nearly 100 years ago, but it is only in the past few years that the development of the technique and associated instrumentation has reached a level that might be called the '*steady state*'. The process and methodology is now well established, and understood, and the practice of the technique is no longer the exclusive domain of the chromatographer. Chromatography is now part of the armory of techniques used by the general analyst and the chromatograph is now just another instrument among the many that are present in the analytical laboratory. The technique of chromatography, however, although well established, is not simple. In the same way that it is necessary for the analyst to have a basic understanding of infra-red spectroscopy in order to use an infra-red spectrometer effectively for analytical purposes, so must the analyst have a basic understanding of the chromatographic process to use the chromatograph effectively. It is the purpose of this book to provide the analyst with a clear, basic understanding of the chromatographic process, the specifications of the equipment necessary to carry out a chromatographic analysis and the procedures that are essential to ensure that the technique is used efficiently.

A Short History of LC

Chromatography was invented by a Russian botanist named Tswett somewhere around the turn of the last century. Tswett, in fact, was educated in Switzerland but returned to Russia to carry out research on plant pigments. His technique was to allow a plant extract to percolate through a bed of powdered calcium carbonate. He reported his findings at the Biological Section of the Warsaw Naturalist's Society in 1903 (1). The colored bands produced on the adsorbent inspired the term chromatography to describe the separation process, combining the Greek word *chromos* meaning color with *grafe* meaning writing. Although color has little to do with modern chromatography, the name has persisted and, despite its irrelevance, is still used to describe all separation techniques that employ a mobile and stationary phase. Unfortunately, the work of Tswett was not immediately developed to any significant extent; this was due in part to the obscurity of the source of the original paper and to the condemnation of the method by Willstatter and Stoll (2). Willstatter and Stoll tried to repeat the work of Tswett but did not heed the advice of Tswett to avoid "aggressive" adsorbents and so their experiments failed. Inaccurate and careless experimental work has always been a threat to scientific progress and recent years have shown that contemporary science is by no means immune to such threats. However, the mistake of Willstatter and Stoll was particularly unfortunate as, not only did it retard the development of a very useful separation technique, but in doing so, it also inhibited progress in many other fields of chemistry.

A further thirty years were to pass before Kuhn and his co-workers (3) successfully repeated Tswett's original work and separated lutein and xanthine from a plant extract. Nevertheless, despite the success of Kuhn *et al* and the validation of Tswett's experiments, the new technique attracted little interest and progress continued to be slow and desultory. In 1941 Martin and Synge (4) introduced liquid-liquid chromatography by supporting the stationary phase, in this case water, on silica in the form of a packed bed and used it to separate some acetyl amino acids.

It is interesting to note that although the renaissance of LC began in 1963 and, in fact, has only recently matured to the level where high resolving power and fast separations are attainable, the essential requirements for HPLC (High Performance Liquid Chromatography) were unambiguously stated by Martin in 1941. To quote Martin's original paper,

"*Thus, the smallest H.E.T.P.* (the highest efficiency) *should be obtainable by using very small particles and a high pressure difference across the column*".

The statement made by Martin in *1941* contains all the necessary conditions to realize both the high efficiencies and high resolution achieved by modern LC columns. Despite his recommendations, however, it has taken nearly fifty years to bring his concepts to fruition. In the same paper Martin and Synge suggested that it would be advantageous to replace the liquid mobile phase by a gas to improve the rate of transfer between the phases and thus, enhance the separation. The recommendation was not heeded and it was left to James and Martin (5) to bring the concept to practical reality in the early fifties. Thus, gas chromatography (GC) was born and a new and important era of chromatography development began.

Gas chromatography grew from a laboratory novelty into a popular well-established analytical technique in little more than a decade. During that time, the foundations of chromatography *column theory* were laid down which were to guide the development of LC in the sixties and seventies. As a result of the glamorous success of GC, LC became the Cinderella of chromatography and it was not until the major developments of GC were completed that scientists turned their attention once again to the development of LC. In contrast to GC, however, the progress of LC has been slow and arduous. The difficulties encountered in the development of LC arose from two causes. Firstly, substances have a very low diffusivity in liquids compared with that of a gas and thus, the kinetics of exchange between the phases are slow and special steps must be taken to achieve efficient separations. Secondly, very low concentrations of a solute in a liquid do not modify the properties of the liquid to the same extent that they do to those of a gas and this has made the development of LC detectors far more difficult. Today, LC is a well-established separation technique, a technique which, to the surprise of many, has established a far wider field of application than GC. It is now a technique with which no analyst can afford to be unfamiliar.

The Separation Process

Chromatography has been defined in the classical manner as

"A separation process that is achieved by the distribution of substances between two phases, a stationary phase and a mobile phase. Those solutes distributed preferentially in the mobile phase will move more rapidly through the system than those distributed preferentially in the stationary phase. Thus, the solutes will elute in order of their increasing distribution coefficients with respect to the stationary phase".

This definition is a little trite and, although it introduces the concept of a *mobile* and *stationary* phase which are essential characteristics of a chromatographic separation, it tends to obscure the basic process of retention in the term *distribution*. A solute is distributed between two phases as a result of the molecular forces that exist between the solute molecules and those of the two phases. The stronger the forces between the solute molecules and those of the stationary phase, the greater will be the amount of solute held in the stationary phase under equilibrium conditions. Conversely, the stronger the interactions between the solute molecules and those of the mobile phase then the greater the amount of solute that will be held in the mobile phase. Now, *the solute can only move through the chromatographic system while it is in the mobile phase* and thus, the speed at which a particular solute will pass through the column will be directly proportional to the concentration of the solute in the mobile phase.

The concentration of the solute in the mobile phase is inversely proportional to the distribution coefficient of the solute *with respect to the stationary phase*.

That is,

$$K = \frac{X_S}{X_m} \tag{1}$$

where (K) is the distribution coefficient of the solute between the two phases,
(X_m) is the concentration of the solute in the mobile phase,
and (X_S) is the concentration of the solute in the stationary phase.

Consequently, the solutes will pass through the chromatographic system at speeds that are inversely proportional to their distribution coefficients with respect to the stationary phase. The control of solute retention by the magnitude of the solute distribution coefficient will be discussed in the next chapter.

To appreciate the nature of a chromatographic separation, the manner of solute migration through a chromatographic column needs to be understood. Consider the progress of a solute through a chromatographic column as depicted in diagrammatic form in figure 1.

Figure 1

The Passage of a Solute Band Along a Chromatographic Column

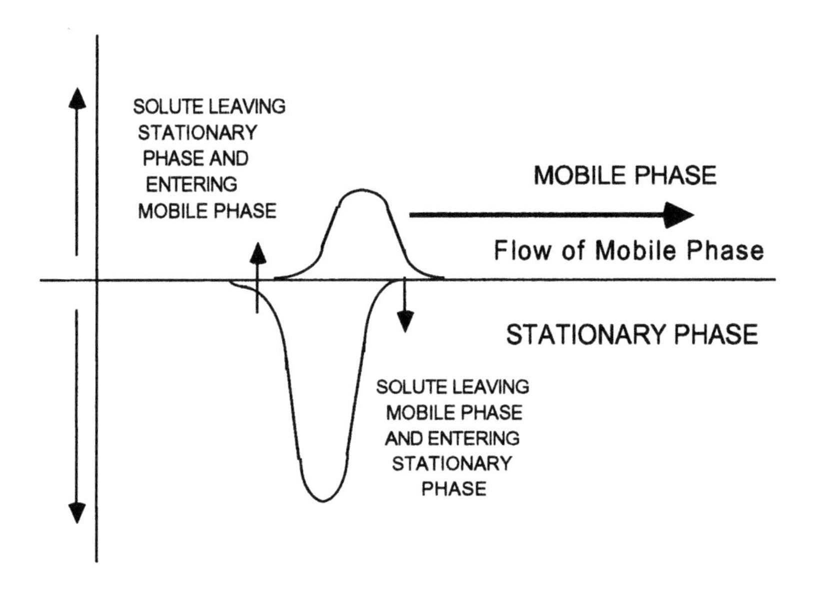

The profile of the concentration of a solute in both the mobile and stationary phases is Gaussian in form and this will be shown to be true when dealing later with basic chromatography column theory. Thus, the flow of mobile phase will slightly displace the concentration profile of the solute in the mobile phase relative to that in the stationary phase; the displacement depicted in figure 1 is grossly exaggerated to demonstrate this effect. It is seen that, as a result of this displacement, the concentration of solute in the *mobile phase* at the front of the peak *exceeds* the equilibrium concentration with respect to that in the stationary phase. It follows that there is a net transfer of solute from the mobile phase in the front part of the peak to the stationary phase to re-establish equilibrium as the peak progresses along the column. At the rear of the peak, the converse occurs. As the concentration profile moves forward, the concentration of solute in the *stationary phase* at the rear of the peak is now in *excess* of the equilibrium concentration. Thus, solute *leaves* the stationary phase and there is a net transfer of solute to the mobile phase in an attempt to re-establish equilibrium. The solute band progresses through the column by a net transfer of solute to the mobile phase at the rear of the peak and a net transfer of solute to the stationary phase at the front of the peak.

Solute retention, and consequently chromatographic resolution, is determined by the magnitude of the distribution coefficients of the solutes with respect to the stationary phase and relative to each other. As already suggested, the magnitude of the distribution coefficient is, in turn, controlled by molecular forces between the solutes and the two phases. The procedure by which the analyst can manipulate the solute/phase interactions to effect the desired resolution will also be discussed in chapter 2.

The Different Forms of Chromatography

The two major classes of chromatography are defined by the nature of the mobile phase. *Gas chromatography* (GC) is the term given to all separations that employ a gas as the mobile phase. Conversely, all chromatographic systems that employ a liquid as the mobile phase are classed as *liquid chromatography*. It has been suggested that *supercritical chromatography*, where, during the separation, the mobile phase can be operated partly above its critical state and partly below its critical state, represents a third class of chromatography. In fact, it is a hybrid: when the mobile phase is below its critical temperature and above its critical pressure, it is a liquid and therefore a liquid chromatography system. When the mobile phase is above its critical temperature and below its critical pressure, it is a gas and therefore the system can be classed as gas chromatography.

It is interesting to note that there have been very few specific analytical applications reported, which demonstrate that super-critical chromatography gives results superior to either gas chromatography or liquid chromatography alone. As a consequence, considering the added complexity of the super-critical chromatograph, its value in analytical chemistry must be considered questionable.

Table 1

Classification of the Different Forms of Chromatography

MOBILE PHASE	STATIONARY PHASE
GAS GAS CHROMATOGRAPHY (GC)	LIQUID GAS-LIQUID CHROMATOGRAPHY (GLC)
	SOLID GAS-SOLID CHROMATOGRAPHY (GSC)
LIQUID LIQUID CHROMATOGRAPHY (LC)	LIQUID LIQUID-LIQUID CHROMATOGRAPHY (LLC)
	SOLID LIQUID-SOLID CHROMATOGRAPHY (LSC)

Each of the major classes of chromatography, gas chromatography and liquid chromatography gives rise to two further classes of chromatography that are defined by the nature of the stationary phase. If the mobile phase is a gas and the stationary phase a liquid, then the system is termed *gas-liquid chromatography* (GLC); conversely if the stationary phase is a solid then this gives rise to *gas-solid chromatography* (GSC). In a similar manner there can be two forms of liquid chromatography, *liquid-liquid chromatography* (LLC), where the stationary phase is a liquid, and *liquid-solid chromatography* (LSC), where the stationary phase is a solid.

The vast majority of modern liquid chromatography systems involve the use of silica gel or a derivative of silica gel, such as a bonded phase, as a stationary phase. Thus, it would appear that most LC separations are carried out by liquid-solid chromatography. Owing to the adsorption of solvent on the surface of both silica and bonded phases, however, the physical chemical characteristics of the separation are more akin to a liquid-liquid distribution system than that of a liquid-solid system. As a consequence, although most modern stationary phases are in fact solids, solute distribution is usually treated theoretically as a liquid-liquid system.

There are other forms of chromatography classification that have been proposed based on the geometric form of the distribution system. For example, the terms column chromatography and lamina chromatography attempt to differentiate between chromatography systems that employ a column and those that employ a sheet of paper, as in paper chromatography, or a sheet of glass coated with silica, as in thin-layer chromatography. Obviously the distribution system involved in paper chromatography is the stationary water or solvent held on the paper and the solvent or solvent mixture that passes over the surface. Such a distribution system is unambiguously liquid-liquid. The silica gel or bonded phase on the thin layer plate however, as in a LC column, is covered by an adsorbed layer of solvent and, consequently, also behaves as a liquid-liquid system. In general, it is better to classify the type of chromatography on the basic physical chemistry involved than, somewhat arbitrarily, on the shape of the apparatus. As a consequence, classification on the basis of the geometry of the distribution system is less commonly employed and, with alternative well-established terms available, the term lamina chromatography is hardly justifiable.

Chromatography Nomenclature

Before proceeding to a more detailed discussion of LC apparatus or separation technology, it is necessary to define some chromatographic terms and, in particular, the properties of a chromatogram that will be used throughout this book. The terms that will be defined were introduced many years ago and have been agreed upon by a number of organizations and are now in common use in all aspects of separation science.

Figure 2

The Nomenclature of a Chromatogram

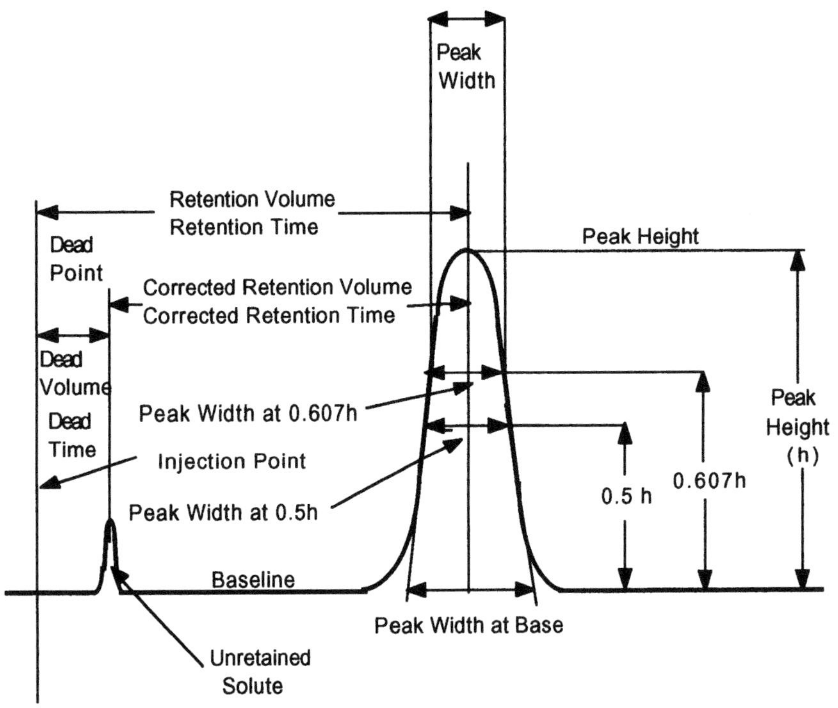

Some early sources of chromatography nomenclature are given in references (6) and (7). The nomenclature of a chromatogram is shown in figure 2, which depicts a chromatogram showing the elution of a single solute. There are three important locations on the chromatogram. Firstly, there is the injection point that locates the start of the chromatogram and, incidentally, the start of the chromatographic development.

Secondly, there is the dead point that identifies the position of the elution of an *unretained* solute and, thirdly, there is the maximum of the solute profile that identifies the position of the peak. The flat portion of the chromatogram where there is no solute being eluted is called the *baseline*. The baseline should be straight and unperturbed but at high sensitivities (usually the two highest sensitivity settings on the detector amplifier) there may be some high frequency excursions of the trace (about 1-2% full scale deflection (FSD)).

The *dead point* is obtained by including in the sample a trace of an unretained solute or, more often, one of the components of the mobile phase. For example, when using a methanol water mixture as the mobile phase, the dead point is obtained from the elution of a pure sample of methanol. The pure methanol can often be monitored, even by a UV detector, as the transient change in refractive index resulting from the methanol is sufficient to cause a disturbance that is detectable.

The time elapsed between the injection and the elution of the unretained solute is called the *dead time* and has been given the symbol (t_o). The volume of mobile phase that passes through the column during the time (t_o) is called the *dead volume* (V_o) where

$$V_o = Qt_o$$

and (Q) is volume flow of mobile phase through the column in ml/min.

The time elapsed between the injection point and the peak maximum is called the *retention time* (t_r) and the volume of mobile phase that passes through the column between the injection and the peak maximum is called the *retention volume* (V_r).

Consequently,

$$V_r = Qt_r$$

In a similar manner the elapsed time between the elution of the unretained solute and the peak maximum is called the *corrected retention time* (t'_r). The volume of mobile phase that passes through the column between the elution of the unretained peak and the solute of interest is called the *corrected retention volume* (V'_r).

Thus, $$V'_r = V_r - V_o = Q(t_r - t_o)$$

The corrected retention volume and corrected retention time can be used for solute identification but more appropriate measurements for this purpose will be discussed later.

The *peak height* is taken as the distance between the extended base line beneath the peak and the peak maximum. The peak height, under certain conditions, will be proportional to the mass of solute present in the peak and can, thus, be used in quantitative analysis. However, the most common measurement employed in quantitative analysis is the *peak area*.

There are a number of values for the peak width that have evolved over the years. The first, called the *peak width at the base*, is obtained from the distance between the points of intersection of the tangents drawn to the sides of the peak with the base line produced beneath the peak. This distance can be shown to be equivalent to four standard deviations of the peak, i.e. (4σ), assuming the peak is not overloaded and is Gaussian in form. The most useful measurement of peak width is that taken at the points of inflection of the Gaussian curve and represents two standard deviations of the peak, i.e. (2σ). This, from a practical point of view, can be shown to be the peak width at 0.6065 of the peak height but is given the general term of *peak width*. Unless otherwise defined, when used throughout this book, the term *peak width* will always refer to the peak width at 0.6065 of the peak height. Another value for the peak width that is sometimes used is the *peak width at half height.* The method of measurement is self-explanatory and is claimed to be a simpler method of measurement. However, whereas the width at a 0.6065 of the peak height has a theoretical significance, the width at half height does not and is, therefore, not recommended for use in column design or column evaluation.

In the early days of liquid chromatography the above measurements were made manually on the chart provided by the potentiometric recorder. Even today a large number of LC systems are simple in form and utilize the potentiometric recorder for monitoring the separation. Accurate results can be obtained from such systems and they certainly cost considerably less than their more sophisticated counterparts. However, many modern and more expensive liquid chromatographs have dispensed with the potentiometric recorder and have associated computer data acquisition and data processing systems which in turn pass information to laboratory management computers. As a consequence, all the measurements described above are calculated from the raw data by the computer employing appropriate software and the results presented in the form of a printed report. Nevertheless, when involved in column design, or procedures where the appropriate software is not available, then the analyst may need to resort to manual measurement and calculation to obtain the required information.

References

1/ M.S.Tswett, *Tr. Protok. Varshav. Obshch. Estestvoispyt Otd. Biol.* **14**(1905).

2/ R. Willstatter and A.Stoll, *Utersuchungenuber Chlorophy,* Springer, Berlin, 1913.

3/ R.Kuhn,A.Winterstein and E.Lederer, *Hoppe-Seyler's Z. Physiol. Chem.*,**197**(1931)141.

4/ A.J.P. Martin and R.L.M. Synge,*Biochem. J.* , **35**(1941)1358.

5/ A.T. James and A.J.P. Martin,*Biochem. J.* , **48**(1951)vii.

6/ R.G.Primavesi, *J. Inst. P.* **527**(1967)367.

7/ R.G.Primavesi, *Pure and Appl. Chem.***No.1**(1960)177.

2
Resolution, Retention and Selectivity

The separation of a mixture into its individual components takes place in the chromatographic column. The column is a simple tube, a few centimeters long and a few millimeters in diameter, packed with particulate material through which the mobile phase permeates. Despite the apparent complexity of the modern chromatograph, the separation is completed in this very simple device, the remaining apparatus providing solvent storage, control of the mobile phase composition, adequate mobile phase inlet pressure, sampling facilities, detection and data processing. As a result of modern instrument design, the essential nature of the column is frequently lost in the electronic and engineering intricacy of the overall system. Consequently, the importance of the column and its function may only be superficially understood, and the technique is not employed in the most efficient manner.

The chromatographic column has a dichotomy of purpose. During a separation, two processes ensue in the column, continuously, progressively and virtually independent of one another. Firstly, the individual solutes are moved apart as a result of the differing distribution coefficients of each component with respect to the stationary phase in the manner previously described. Secondly, having moved the individual components apart, the column is designed to constrain the natural dispersion of each solute band (i.e. the band spreading) so that, having been separated, the components are eluted discretely from the column as *individual* solute bands.

Thus, the column ***moves*** *the solutes bands apart and simultaneously* ***contains*** *their dispersion .*

As the individual components of a mixture are moved apart on the basis of their differing retention, then the separation can be partly controlled by the choice of the phase system. In contrast, the peak dispersion that takes place in a column results from kinetic effects and thus is largely determined by the physical properties of the column and its contents.

Thus, for optimum performance, each sample will require a specific phase system to be chosen and a particular column to be selected .

In order to understand the mechanism of retention and selectivity and thus be in a position to exercise some control over the chromatographic system, it is necessary to derive an equation for the retention volume of a solute. Such an equation would enable the analyst to further understand the process of retention and also identify the pertinent variables that need to be controlled to carry out a specific analysis. To employ liquid chromatography in a satisfactory manner, it is *not* necessary to have a profound understanding of liquid chromatography column theory. It is, however, advisable to have a basic knowledge of the subject to understand the rationale behind phase selection and column design. In this book, chromatography theory will only receive minimal treatment but it will include those aspects essential to the successful operation of the liquid chromatograph. Those interested in pursuing the subject further are recommended to read "Liquid Chromatography Column Theory " published by Wiley (1). Now, in order to obtain an equation for the retention volume of a solute, a basic measurement by which solute identification is accomplished, it is necessary to discuss the Plate Theory.

The Plate Theory

Primarily, the Plate Theory provides the equation for the elution curve (the chromatogram) of a solute in terms of the volume of mobile phase that has passed through it. From this equation, the various characteristics of a chromatographic system can be determined using the data that is provided by the chromatogram. The Plate Theory was originally derived by Martin and Synge (2) and was based on the 'plate' concept employed in the theory of distillation columns. The theory of Martin and Synge was later modified by Said (3,4) and it is the procedure of Said that will be given here. The Plate Theory assumes that the solute is, at all times, in equilibrium with both the mobile and stationary phases. However, due to the continuous exchange of solute between the two phases as it progresses through the column, equilibrium between the phases is, in fact, never actually achieved. As a consequence, the column is considered to be divided into a number of theoretical plates. Each plate is allotted a finite length, and thus, the solute is considered to spend a finite time in each plate. The size of the plate is such that the solute is assumed to have sufficient time to achieve equilibrium with the two phases. It follows that if the exchange is fast and efficient, the theoretical plate will be small in size and there will be a large number of plates in the column. Conversely, if the column performance is poor, the exchange of solute between the phases will be slow and the theoretical plates larger and fewer in number.

Consider the equilibrium existing in each plate, then

$$X_S = KX_m \qquad (1)$$

where (X_m) and (X_S) are, respectively, the concentrations of the solute in the mobile and stationary phases, and (K) is the distribution coefficient of the solute between the two phases. (It should be remembered that the distribution coefficient is defined with reference to the stationary phase, i.e. $K = X_S/X_m$, and thus the larger the distribution coefficient, the greater the proportion of solute that is distributed in the stationary phase.)

Equation (1) merely states that the general distribution law applies to the system and that the adsorption isotherm is linear. At the concentrations normally employed in liquid chromatographic separations this will be true.

Differentiating equation (1),

$$dX_S = K dX_m \qquad (2)$$

Consider three consecutive plates in a column, the (p-1), the (p) and the (p+1) plates and let there be a total of (n) plates in the column. The three plates are depicted in figure 1.

Figure 1

Three Consecutive Theoretical Plates in an LC Column

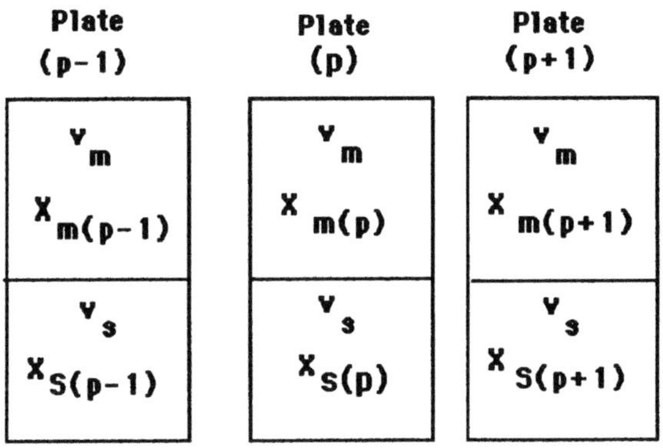

Plate (p-1)	Plate (p)	Plate (p+1)
v_m	v_m	v_m
$X_{m(p-1)}$	$X_{m(p)}$	$X_{m(p+1)}$
v_s	v_s	v_s
$X_{S(p-1)}$	$X_{s(p)}$	$X_{S(p+1)}$

Let the volumes of mobile phase and stationary phase in each plate be (v_m) and (v_s) respectively and the concentrations of solute in the mobile and stationary phase in each plate be $X_{m(p-1)}$, $X_{s(p-1)}$, $X_{m(p)}$, $X_{s(p)}$, $X_{m(p+1)}$, and $X_{s(p+1)}$ respectively. Let a volume of mobile phase, (dV), pass from plate (p-1) into plate (p) at the same time displacing the same volume of mobile phase from plate (p) to plate (p+1). As a consequence, there will be a change of mass of solute in plate (p) that will be equal to the difference in the mass entering plate (p) from plate (p-1) and the mass of solute leaving plate (p) and entering plate (p+1).

Thus, bearing in mind that mass is the product of concentration and volume, the change of mass of solute in plate (p) is:

$$dm = \left(X_{m(p-1)} - X_{m(p)}\right)dV \tag{3}$$

Now, if equilibrium is to be maintained in the plate (p), the mass (dm) will distribute itself between the two phases, which will result in a change of solute concentration in the mobile phase of $dX_{m(p)}$ and in the stationary phase of $dX_{s(p)}$.

Thus,
$$dm = v_s dX_{s(p)} + v_m dX_{m(p)} \tag{4}$$

Substituting for $dX_{s(p)}$ from equation (2),

$$dm = (v_m + Kv_s)dX_{m(p)} \tag{5}$$

Equating equations (3) and (5) and re-arranging,

$$\frac{dX_{m(p)}}{dV} = \frac{X_{m(p-1)} - X_{(p)}}{v_m + Kv_s} \tag{6}$$

Now, to aid in algebraic manipulation the volume flow of mobile phase will now be measured in units of $(v_m + Kv_s)$ instead of milliliters. Thus the new variable (v) can be defined where

$$v = \frac{V}{(v_m + Kv_s)} \tag{7}$$

The function $(v_m + Kv_s)$ has been given the name 'plate volume' and thus, for the present, the flow of mobile phase through the column will be measured in 'plate volumes' instead of milliliters.

Differentiating equation (7),

$$dv = \frac{dV}{(v_m + Kv_s)} \qquad (8)$$

Substituting for dV from (8) in (6)

$$\frac{dX_{m(p)}}{dv} = X_{m(p-1)} - X_{(p)} \qquad (9)$$

Equation (9) is the basic differential equation that describes the rate of change of concentration of solute in the mobile phase in plate (p) with the volume flow of mobile phase through it. The integration of equation (9) will provide the equation for the elution curve of a solute for any plate in the column. A detailed integration of equation (9) will not be given here and the interested reader is again directed to reference (1) for further details.

Integrating equation (9),

$$X_{m(n)} = \frac{X_o e^{-v} v^n}{n!} \qquad (10)$$

where $X_{m(n)}$ is the concentration of solute in the mobile phase leaving the (n)th plate (i.e. the concentration of solute entering the detector)

and X_o is the initial concentration of solute on the 1st plate of the column.

Equation (10) describes the elution curve obtained from a chromatographic column and is the equation of the curve, or chromatogram, that is traced by the chart recorder or computer printer. Its pertinence is displayed in figure 2.

Figure 2

The Elution of a Single Solute

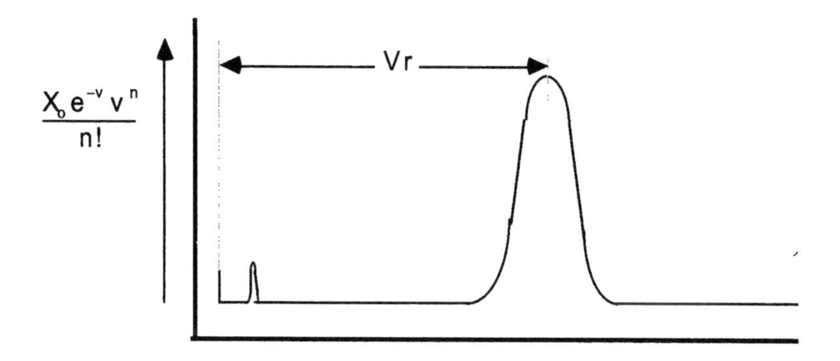

It can now be seen how an expression for the retention volume of a solute can be derived. By differentiating equation (10) and equating to zero, an expression for the volume of mobile phase passed through the column between the injection point and the peak maximum can be obtained. This volume has already been defined as the retention volume (V_r) of the solute.

The Retention Volume of a Solute

Restating equation (10),

$$X_{m(n)} = X_0 \frac{e^{-v} v^n}{n!}$$

$$\frac{dX_{m(n)}}{dv} = X_0 \frac{-e^{-v} v^n + e^{-v} n v^{(n-1)}}{n!}$$

$$= X_0 \frac{-e^{-v} v^{(n-1)}}{n!} (n - v)$$

Equating to zero,

$$n - v = 0$$

or

$$v = n$$

This means that at the peak maximum, (n) plate volumes of mobile phase have passed through the column. Remembering that the volume flow is measured in 'plate volumes' and not milliliters, the volume passed through the column in ml will be obtained by multiplying by the 'plate volume' $(v_m + Kv_s)$.

Thus, the retention volume (V_r) is given by:

$$V_r = n(v_m + Kv_s)$$

$$= nv_m + nKv_s$$

Now the total volume of mobile phase in the column, (V_m), will be the volume of mobile phase per plate multiplied by the number of plates, i.e. (nv_m). In a similar manner the total volume of stationary phase in the column (V_s) will be the volume of stationary phase per plate multiplied by the total number of plates, i.e. (nv_s).

Thus,

$$\mathbf{V_r = V_m + KV_s} \tag{11}$$

It is now immediately obvious from equation (11) how the separation of two solutes (A) and (B) must be achieved.

For separation $V_{r(A)} <> V_{r(B)}$ and $V_{r(A)} \neq V_{r(B)}$

Furthermore, as the retention volumes of each substance must be different then,

either $K_{(A)} <> K_{(B)}$ and $K_{(A)} \neq K_{(B)}$

or $V_{S(A)} <> V_{S(B)}$ and $V_{S(A)} \neq V_{S(B)}$

Thus, to achieve the required separation either (1) the distribution coefficient (K) of all the solutes must be made to differ, or (2) the amount of stationary phase (V_S) available to each component of the mixture interacts must be made to differ. A further alternative (3) would be to make appropriate adjustments to both the values of (K) and the values (V_S).

The analyst is now required to know how to change (K) and (V_S) to suit the particular sample of interest. Consider first the control of the distribution coefficient (K).

Factors that Control the Distribution Coefficient of a Solute

Molecular Interactions

The effect of molecular interactions on the distribution coefficient of a solute has already been mentioned in Chapter 1. Molecular interactions are the direct effect of intermolecular forces between the solute and solvent molecules and the nature of these molecular forces will now be discussed in some detail. There are basically four types of molecular forces that can control the distribution coefficient of a solute between two phases. They are *chemical forces*, *ionic forces*, *polar forces* and *dispersive forces*. *Hydrogen bonding* is another type of molecular force that has been proposed, but for simplicity in this discussion, hydrogen bonding will be considered as the result of very strong polar forces. *These four types of molecular forces that can occur between the solute and the two phases are those that the analyst must modify by choice of the phase system to achieve the necessary separation.* Consequently, each type of molecular force enjoins some discussion.

Chemical Forces

Chemical forces are normally irreversible in nature (at least in chromatography) and thus, the distribution coefficient of the solute with respect to the stationary phase is infinite or close to infinite. Affinity chromatography is an example of the use of chemical forces in a separation process. The stationary phase is formed in such a manner that it will chemically interact with one unique solute present in the sample and thus, exclusively extract it from the other materials present. The technique of affinity chromatography is, therefore, an extraction process more than a chromatographic separation, an aspect of separation science that is not germane to the subject of this book; *ipso facto,* the nature of chemical forces need not be discussed further.

Ionic Forces

Ionic forces are electrical in nature and result from the charges produced when molecules ionize in solution into positively charged cations and negatively charged anions. The resulting ionic interactions are exploited in ion chromatography. For example in the analysis of organic acids, it is the negatively charged acid anions that require to be separated. The stationary phase must, therefore, contain positively charged cations as counter ions to interact with the acid anions, retard them in the column and effect their resolution. Conversely, to separate cations, the stationary phase must contain anions as counter ions with which the cations can interact.

Ion exchange stationary phases are usually available in the form of cross-linked polymer beads that have been appropriately modified to contain the desired ion exchange groups. The material is supplied in a form ready for packing or in pre-packed columns. Alternatively, ion exchange groups can be chemically bonded to silica by a process similar to the preparation of ordinary bonded phases. Bonded phases and their production will be discussed in a later chapter. Ion exchange materials can also be *adsorbed* on the surface of a bonded phase which can then act as an adsorbed ion exchanger. Such materials are usually called *Ion Pair Reagents* and are added to the mobile phase in relatively low concentrations (ca. 1%w/v) from which, under conditions of low solvent concentration, they can be absorbed onto the surface of the stationary phase. Examples of the use ion pair reagents will also be discussed in thc chapter dealing specifically with stationary phases. Examples of the separation of a series of inorganic anions on an ion exchange column is shown in figure 3.

Figure 3

The Separation of a Series of Inorganic Anions

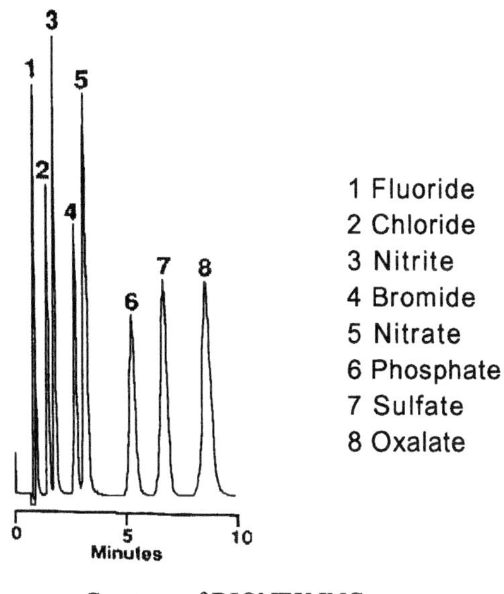

Courtesy of DIONEX INC.

The column was 15 cm long, 4.6 mm in diameter, and packed with a proprietary ion exchange material IonPacAS4A. The mobile phase was an aqueous solution of 1.80 nM sodium carbonate and 1.70 nM sodium bicarbonate at a flow rate of 2.0 ml/min. The volume of charge was 50 μl.

It should be emphasized that in figure 3, the system was not operated at its maximum sensitivity and, at higher sensitivity settings, it could be employed directly for the analysis of drinking water and other types of water samples. Further practical information can be obtained from the books by Small (5), Weiss (6) and Smith (7).

Polar Forces

Polar forces also arise from electrical charges on the molecule but in this case from permanent or induced dipoles. It must be emphasized that there is *no net charge* on a polar molecule unless it is *also* ionic. Examples of substances with permanent dipoles are alcohols, esters, aldehydes etc. Examples of polarizable molecules are the aromatic hydrocarbons such as benzene, toluene and xylene. Some molecules can have permanent dipoles and, at the same time, also be polarizable. An example of such a substance would be phenyl ethyl alcohol. The hydroxyl group of the alcohol would form part of a permanent dipole whereas the phenyl group would be polarizable. To separate polar materials on a basis of their polarity, a *polar* stationary phase must be used. Furthermore, in order to focus the polar forces in the stationary phase, and consequently the selectivity, a relatively *non-polar* mobile phase would be required.

The use of dissimilar molecular interactions in the two phases, to achieve selectivity, is generally applicable and of fundamental importance.

The specific interactions that will produce the necessary retention and selectivitv must dominate in the stationary phase to achieve the separation. It follows that it is important that they are also as *exclusive as possible* to the stationary phase. It is equally important to ensure that the interactions taking place in the mobile phase *differ to as great extent as possible* to that in the stationary phase in order to maintain the stationary phase selectivity.

To separate some aromatic hydrocarbons, for example, a strongly polar stationary phase would be appropriate as the solutes do not contain permanent dipoles. Under such circumstances, when the polarizable aromatic nucleus approaches the strongly polar group on the stationary phase, the aromatic nucleus will be polarized and positive and negative charges generated at different positions on the aromatic ring. These charges will then interact with the charges from the dipole of the stationary phase and, as a consequence, the solute molecules held more strongly and distributed more favorably in the stationary phase. In practice, *silica gel* is a strongly polar stationary phase, its polar character arising from the dipoles of the surface hydroxyl groups. Consequently, silica gel would be an appropriate stationary phase for the separation of aromatic hydrocarbons. To ensure that polar selectivity *dominates* in the *stationary phase*, and polar interactions in the mobile phase are minimized, a non-polar or dispersive solvent, e.g. n-heptane, could be used as the mobile phase. The separation of some biphenyl derivatives on silica gel is shown in figure 4.

Figure 4

The Separation of a Mixture of Aromatic and Nitro-Aromatic Hydrocarbons

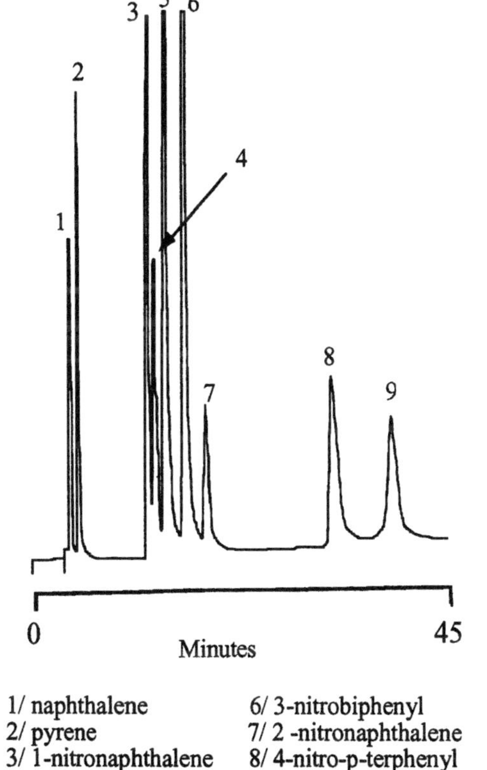

1/ naphthalene	6/ 3-nitrobiphenyl
2/ pyrene	7/ 2 -nitronaphthalene
3/ 1-nitronaphthalene	8/ 4-nitro-p-terphenyl
4/ 4-nitroanthracene	9/ 1-nitropyrene
5/ 9-nitroanthracene	

The separation was carried out on a small bore column 25 cm long and 1 mm i.d. packed with silica gel having a particle diameter of 10μ. The mobile phase was *n*-hexane saturated with water and the flow rate 50 μl per min.

Dispersive Forces

Dispersive forces are more difficult to describe. Although electric in nature, they result from charge fluctuations rather than permanent electrical charges on the molecule. Examples of purely dispersive interactions are the molecular forces that exist between saturated aliphatic hydrocarbon molecules. Saturated aliphatic hydrocarbons are not ionic, have no permanent dipoles and are not polarizable. Yet molecular forces between hydrocarbons are strong and consequently, n-heptane is not a gas, but a liquid that boils at 100°C. This is a result of the collective effect of all the dispersive interactions that hold the molecules together as a liquid.

To retain solutes selectively by dispersive interactions, the stationary phase must contain no polar or ionic substances, but only hydrocarbon-type materials such as the reverse-bonded phases, now so popular in LC. Reiterating the previous argument, to ensure that dispersive selectivity dominates in the stationary phase, and dispersive interactions in the mobile phase are minimized, the mobile phase must now be strongly polar. Hence the use of methanol-water and acetonitrile-water mixtures as mobile phases in reverse-phase chromatography systems. An example of the separation of some anti-microbial agents on Partisil ODS 3, particle diameter 5μ is shown in figure 5.

ODS3 is a "bulk type" reverse phase (the meaning of which will be discussed later) which has a fairly high capacity and is reasonably stable to small changes in pH. The column was 25 cm long, 4.6 mm in diameter and the mobile phase a methanol water mixture containing acetic acid. In this particular separation the solvent mixture was programmed, a development procedure which will also be discussed in a later chapter.

The components present in the mixture are as follows:
a/ sulfaguanidine, b/ sulfanilamide, c/ sulfadiazine, d/ sufathiazole, e/ sulphapyridine, f/ sulfamerazine, g/ sulfamethazine,

h/ sulfachloropyridazine, i/ sulfasoxazole, j/ sulfaethoxypyridazine, k/ sulfadimethoxine, l/ sulfaquinoxine, m/ sulfabromethazine.

Figure 5

The Separation of a Mixture of Anti-Microbial Agents

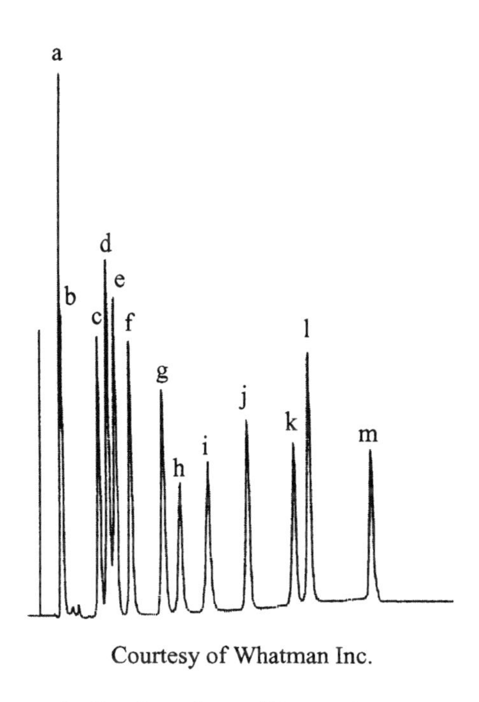

Courtesy of Whatman Inc.

The Thermodynamic Explanation of Retention

Solute retention can also be explained on a thermodynamic basis where the change in free energy is considered when the solute is moved from the environment of one phase to that of the other.

The classic thermodynamic expression for the distribution coefficient (K) of a solute between two phases is given by

$$RT \ln K = -\Delta G_0$$

where (R) is the Gas Constant,
(T) is the absolute temperature,
and (ΔG_0) is the Excess Free Energy.

Now,

$$\Delta G_0 = \Delta H_0 - T\Delta S_0$$

where

(ΔH_0) is the Excess Free Enthalpy

and (ΔS_0) is the Excess Free Entropy.

Thus,

$$\ln K = -\left(\frac{\Delta H_0}{RT} - \frac{\Delta S_0}{R}\right)$$

or

$$K = e^{-\left[\frac{\Delta H_0}{RT} - \frac{\Delta S_0}{R}\right]}$$

It is seen that, if the *excess free entropy* and *excess free enthalpy* of a solute between two phases can be calculated or predicted, then the magnitude of the distribution coefficient (K) and consequently, the retention of a solute can also be predicted. Unfortunately, the thermodynamic properties of a distribution system are *bulk* properties, that include the effect of all the different types of molecular interactions in one measurement. As a result it is difficult, if not impossible, to isolate the individual contributions so that the nature of overall distribution can be estimated or controlled. Attempts have been made to allot specific contributions to the total *excess free energy* of a given solute/phase system that represents particular types of interaction. Although the procedure becomes very complicated, this empirical approach has proved to be of some use to the analyst. It must be said that, despite "the sound and fury" surrounding the subject, thermodynamics has yet to predict a single distribution coefficient from basic physical chemical data. It is for this reason, that the thermodynamic control of the magnitude of the distribution coefficient (K) is only briefly considered in this book. Nevertheless, if sufficient experimental data is available or can be obtained, then empirical equations, similar to those given above, can be used to optimize a particular distribution system once the basic phases have been identified. A range of computer programs, based on this rationale, is available, that purport to carry out optimization for up to three component solvent mixtures. Nevertheless, the appropriate stationary phase is still usually identified from the types of molecular forces that need to be exploited to effect the required separation.

Equation (12) can also be used to identify the type of retention mechanism that is taking place in a particular separation by measuring the retention volume of the solute over a range of temperatures.

Re-arranging equation (12)

$$\log K = -\frac{\Delta H_0}{RT} + \frac{\Delta S_0}{R}$$

Bearing in mind, $V' = KV_S$

$$\log V' = -\frac{\Delta H_0}{RT} + \frac{\Delta S_0}{R} - \log V_s$$

It is seen that a curve relating Log(V') to 1/T should give a straight line the slope of which will be proportional to the *enthalpy* change during solute transfer. In a similar way, the intercept will be related to the *entropy* change during solute transfer and thus, the dominant effects in any distribution system can be identified from such curves. Graphs of Log(V') against 1/T for two different types of distribution systems are given in figure 6. It is seen that distribution system (A) has a large enthalpy value $\left[\frac{\Delta H_0}{RT}\right]_A$ and a low entropy contribution $\left[-\frac{\Delta S_0}{R} - V_S\right]_A$. In fact, the value of (V_S) would be common to both systems and thus, the value of the entropy change will be proportional to changes in $\left[-\frac{\Delta S_0}{R} - V_S\right]_A$. The large value of $\left[\frac{\Delta H_0}{RT}\right]_A$ means that the distribution is predominantly controlled by molecular forces. The solute is preferentially distributed in the stationary phase as a result of the interactions of the solute molecules with those of the stationary phase being much greater than the interactive forces between the solute molecules and those of the mobile phase. Because the change in enthalpy is the major contribution to the change in free energy,

in thermodynamic terms the distribution is said to be "energy driven".

Figure 6

Graph of Log Corrected Retention Volume against the Reciprocal of the Absolute Temperature

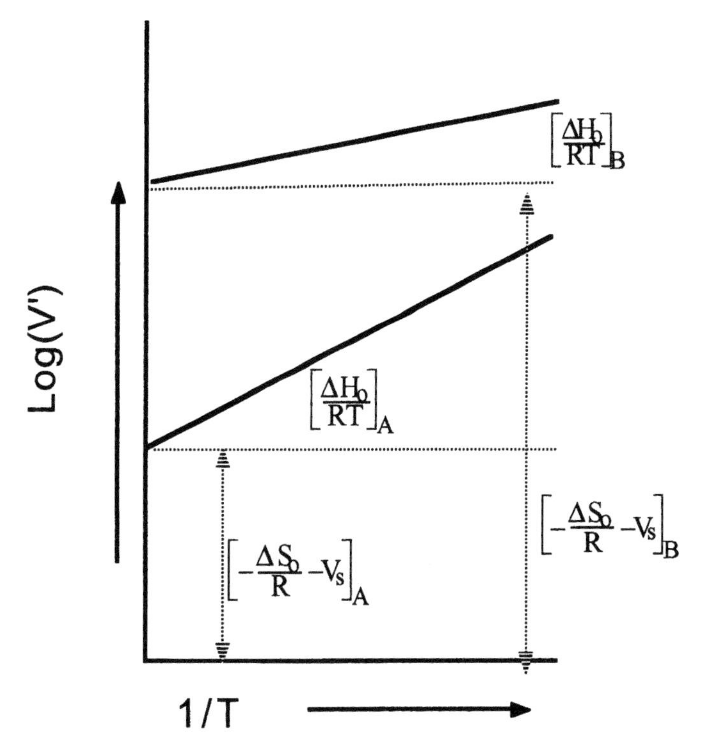

In contrast, it is seen that for distribution system (B) there is only a small enthalpy change $\left[\frac{\Delta H_o}{RT}\right]_B$, but in this case a high entropy contribution $\left[-\frac{\Delta S_o}{R} - V_s\right]_B$. This means that the distribution is *not* predominantly controlled by molecular forces. The entropy change reflects the degree of randomness that a solute molecule experiences in a particular phase. The more random and 'more free' the solute molecule is to move in a particular phase, the greater the entropy of the solute. Now as there is a large entropy change, this means that the solute molecules are more restricted or less random in the stationary phase in system (B). This loss of freedom is responsible for the greater distribution of the solute in the stationary phase and thus, its greater retention. Because the change in entropy in system (B) is the major contribution to the change in free energy,

in thermodynamic terms the distribution is said to be "entropically driven".

Examples of entropically driven systems will be those employing chirally active stationary phases.

Returning to the molecular force concept, in any particular distribution system it is rare that only one type of interaction is present and if this occurs, it will certainly be *dispersive* in nature. Polar interactions are always accompanied by dispersive interactions and ionic interactions will, in all probability, be accompanied by both polar *and* dispersive interactions. However, as shown by equation (11), it is not merely the *magnitude* of the interacting forces between the solute and the stationary phase that will control the extent of retention, but also the *amount of stationary phase* present in the system and its *accessibility* to the solutes. This leads to the next method of retention control, and that is the volume of stationary phase available to the solute.

Factors that Control the Availability of the Stationary Phase

The volume of stationary phase with which the solutes in a mixture can interact (V_S in equation (11)) will depend on the physical nature of the stationary phase or support. If the stationary phase is a porous solid, and the sizes of the pores are commensurate with the molecular diameter of the sample components, then the stationary phase becomes size selective. Some solutes (for example those of small molecular size) can penetrate and interact with more stationary phase than larger molecules which are partially excluded. Under such circumstances the retention is at least partly controlled by size exclusion and, as one might expect, this type of chromatography is called Size Exclusion Chromatography (SEC). Alternatively, if the stationary phase is *chiral* in character, the interaction of a solute molecule with the surface (and consequently, the amount of stationary phase with which in can interact) will depend on the chirallity of the solute molecule and how it fits to the chiral surface. Such separation systems have been informally termed Chiral Liquid Chromatography (CLC). However, it must be pointed out that both SEC and CLC fall within the classes of LLC or LSC depending on the nature of the stationary phase.

Size Exclusion Chromatography

The term size exclusion chromatography implies that the retention of a solute depends solely on solute molecule size. However, even in SEC, retention may not be *exclusively* controlled by the size of the solute molecule, it will still be controlled by molecular interactions between the solute and the two phases. *Only* if the magnitude of the forces between the solute and both phases is the *same* will the retention and selectivity of the chromatographic system depend solely on the pore size distribution of the stationary phase. Under such circumstances the larger molecules, being partially or wholly excluded, will elute first and the smaller molecules elute last. It is interesting to note that, even if the dominant retention mechanism is controlled by molecular forces between the solute and the two phases, if the stationary phase or supporting material has a porosity commensurate with the molecular size of the solutes, *exclusion will still play a part in retention.* The subject of mixed retention mechanisms will be discussed when considering the physical characteristics of silica gel. The two most common exclusion media employed in liquid chromatography are silica gel and macroporous polystyrene divinylbenzene resins. Exclusion chromatograms obtained from three different silica gel columns are shown in figure 7 (8). The three columns, shown in figure 7, were packed with Spherosil 10μ, Silerex 2 10μ and Partisil 10μ, a series of silicas that exhibit quite different exclusion properties.

Figure 7

The Separation of a Mixture by Exclusion Chromatography

Chromatograms Demonstrating the Different Exclusion Properties of Three Silica Gels

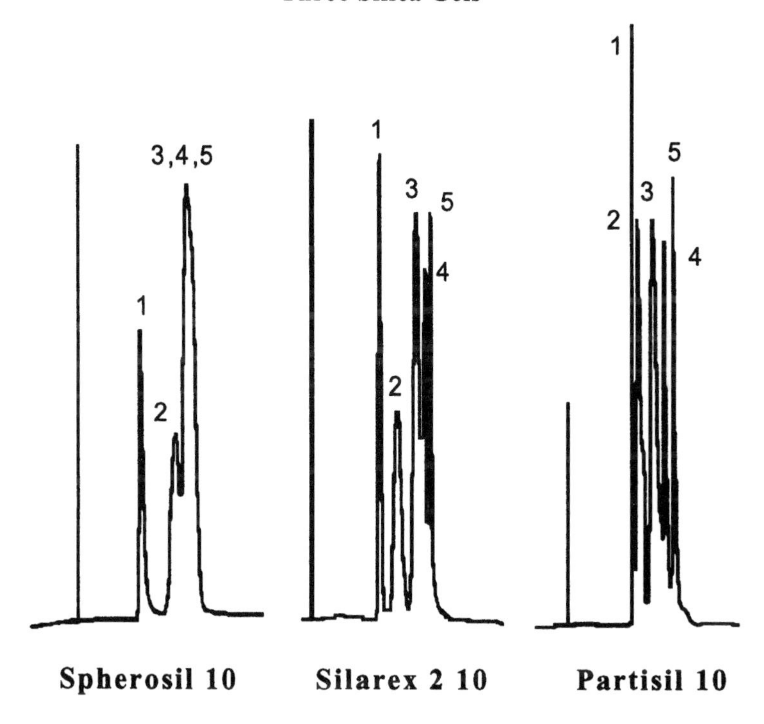

Column Length 50 cm, mobile phase tetrahydrofuran, Molecular diameter of solutes, (1) 11,000Å, (2) 240Å, (3) 49.5Å, (4) 27.1Å and (5) 7.4Å.

The number 10 refers to the diameter of the silica particles in micron and *not* the pore size. The five solutes (1-5) were largely hydrocarbon in nature having mean molecular diameters of 11,000, 240, 49.5, 27.1 and 7.4 Å respectively. The mobile phase employed was tetrahydrofuran (THF). This solvent is adsorbed as a layer on the surface of the silica (a phenomenon that will be discussed in more detail in a later chapter) and thus the solutes interact with THF in *both* phases. Consequently, the distribution coefficient of each solute with respect to the mobile phase is unity and there is no interactive selectivity and the sequence of elution is governed *solely by exclusion*. This is supported by the order of elution of the solutes on each stationary phase shown in figure 7. It is seen that all the *smaller* solutes are bunched together in the chromatogram from Spherosil and thus the material is appropriate for the separation of the larger solutes having average molecular diameters lying between 50 and 1100Å. The chromatogram from Silerex, however, shows a fairly clean separation for all solutes and, therefore, could be used to separate solutes covering the entire molecular diameter range between 7.4Å and 1100Å. At the other extreme, using Partisil, the solutes having molecular diameters of 240Å and 1100Å are eluted first, and close together, whereas, the solutes having diameters ranging from 7.4Å to 49.5 Å are well resolved. It follows that Partisil would be appropriate for the separation of solutes that are relatively small in molecular size. Silica is often used as an exclusion media for the separation of high molecular weight hydrocarbons and polymers whereas polymeric materials of biological origin when separated by exclusion, often employ Sephadex or micro-reticular polystyrene gels as the stationary phase.

Unfortunately, exclusion chromatography has some inherent disadvantages that make its selection as the separation method of choice a little difficult. Although the separation is based on molecular size, which might be considered an ideal rationale, the total separation must be contained in the pore volume of the stationary phase. That is to say all the solutes must be eluted between the excluded volume and the dead volume, which is approximately half the column dead volume. In a 25 cm long, 4.6 mm i.d. column packed with silica gel, this means that all the solutes must be eluted in about 2 ml of mobile phase. It follows, that to achieve a reasonable separation of a multi-component mixture, the peaks must be very narrow and each occupy only a few microliters of mobile phase. Scott and Kucera (9) constructed a column 14 meters long and 1 mm i.d. packed with 5μ silica gel that had an efficiency of 650,000 theoretical plates (probably the highest column efficiency so far produced in HPLC) and used it to separate a mixture of alkyl benzenes by exclusion. The separation obtained by Scott and Kucera is shown in figure 8.

Figure 8

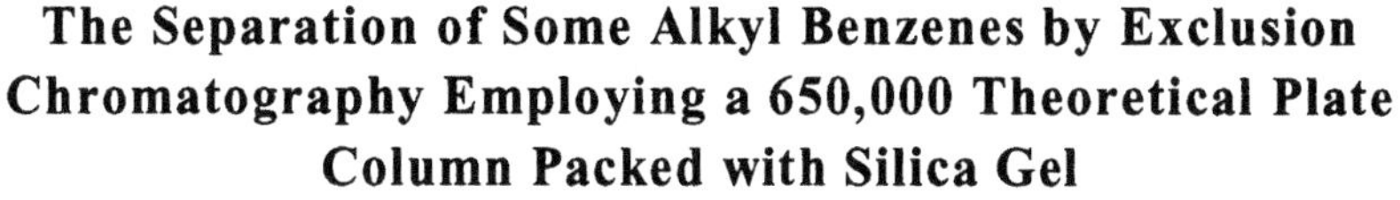

The Separation of Some Alkyl Benzenes by Exclusion Chromatography Employing a 650,000 Theoretical Plate Column Packed with Silica Gel

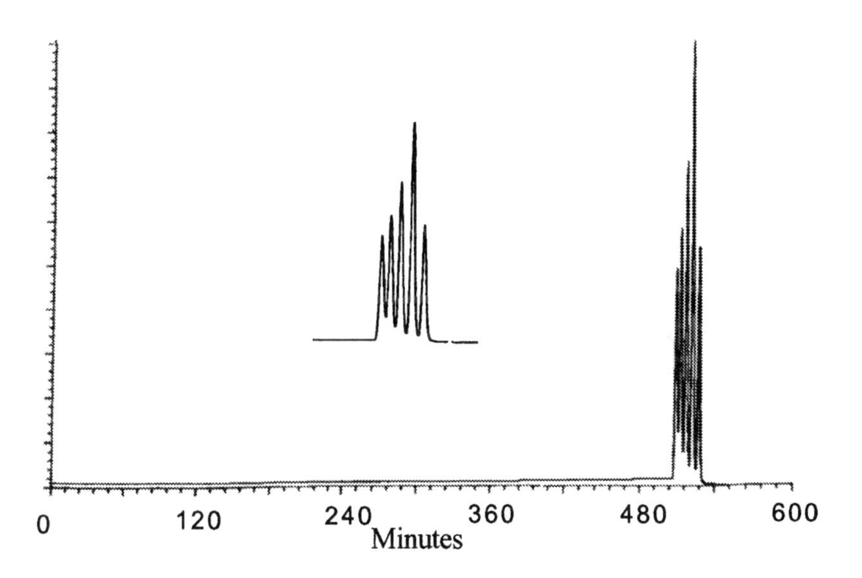

The solutes were benzene, ethyl benzene, butyl benzene, hexyl benzene, octyl benzene and decyl benzene and it is seen that the silica could easily resolve any two solute pairs having a difference of only two methylene groups. In fact, solutes having only one methylene group difference would have been discerned as two peaks. The disadvantage of the system can be seen from the time scale. The separation took over 10 hours to complete and this extensive time period was necessary even when using the maximum inlet pressure available of 6,000 p.s.i.

Nevertheless, despite the inherent disadvantages of exclusion chromatography, there are instances where it is the only practical method of choice. The technique is widely used in the separation of macro-molecules of biological origin, e.g. polypeptides, proteins, enzymes, etc. In fact, it is in this area of biotechnology where the major growth in HPLC techniques appears to be taking place.

Chiral Chromatography

Modern organic chemistry is becoming increasingly involved in methods of asymmetric syntheses. This enthusiasm has been fostered by the relatively recent appreciation of the differing physiological activity that has been shown to exist between the geometric isomers of pharmaceutically active compounds. A sad example lies in the drug, Thalidomide, which was marketed as a racemic mixture of N-phthalylglutamic acid imide. The desired pharmaceutical activity resides in the R-(+)-isomer and it was not found, until too late, that the corresponding S-enantiomer, was teratogenic and its presence in the racemate caused serious fetal malformations. It follows that the separation and identification of isomers can be a very important analytical problem and liquid chromatography can be very effective in the resolution of such mixtures.

Many racemic mixtures can be separated by ordinary reverse phase columns by adding a suitable chiral reagent to the mobile phase. If the material is adsorbed strongly on the stationary phase then selectivity will reside in the stationary phase, if the reagent is predominantly in the mobile phase then the chiral selectivity will remain in the mobile phase. Examples of some suitable additives are camphor sulphonic acid (10) and quinine (11). Chiral selectivity can also be achieved by bonding chirally selective compounds to silica in much the same way as a reverse phase. A example of this type of chiral stationary phase is afforded by the cyclodextrins.

The cyclodextrins are produced by the partial degradation of starch followed by the enzymatic coupling of the glucose units into crystalline, homogeneous toroidal structures of different molecular sizes. Three of the most widely characterized are *alpha, beta* and *gamma* cyclodextrins which contain 6, 7 and 8 glucose units respectively. Cyclodextrins are, consequently, chiral structures and the beta-cyclodextrin has 35 sterogenic centers. CYCLOBOND is a trade name used to describe a series of chemically bonded cyclodextrins to spherical silica gel. The process of bonding is proprietary and patented but a number of CYCLOBOND columns are commercially available. An example of the separation of the isomers of Warfarin is shown in figure 9. The column was 25 cm long and 4.6 mm in diameter packed with 5 micron CYCLOBOND 1. The mobile phase was approximately 90%v/v acetonitrile, 10%v/v methanol, 0.2 %v/v glacial acetic acid and 0.2%v/v triethylamine. It is seen that an excellent separation has been achieved with the two isomers completely resolved.

Figure 9

The Separation of Warfarin Isomers on a CYCLOBOND Column

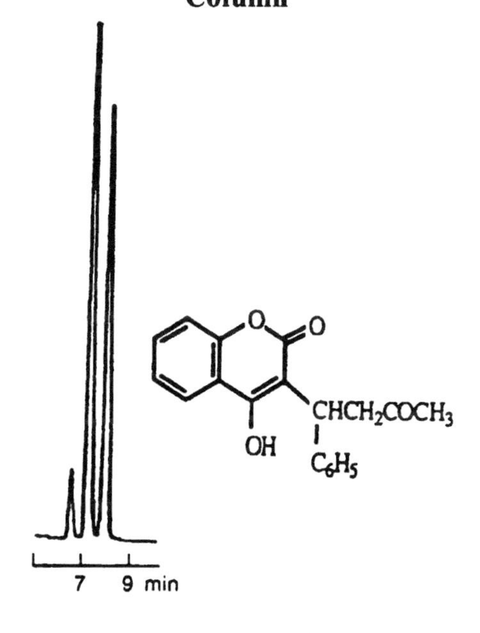

Courtesy of ASTEC Inc.

This separation is an impressive example of an entropically driven distribution system where the normally random movements of the solute molecules are restricted to different extents depending on the spatial orientation of the substituent groups. For further information the reader is directed to an excellent review of chiral separations by LC (Taylor and Maher (12)) and a monograph on CYCLOBOND materials from ASTEC Inc. (13).

Chromatographic Methods of Identification

The first chromatographic parameter that was employed for solute identification was the corrected retention volume (V'_r). However, the measurement of (V'_r) is not as straightforward as it might seem. Consider equation (11), which for convenience will be reiterated.

$$V_r = V_m + KV_S \qquad (11)$$

It must be pointed out that (V_m) refers to the volume of mobile phase *in the column* and not the total volume of mobile phase between the injection valve and the detector (V_o). In practice, the dead volume (V_o) will include all the extra column volumes (V_E) involved in the sample valve, connecting tube detector cell etc.

Thus,
$$V_o = V_m + V_E \qquad (12)$$

In some cases, (V_E) may be sufficiently small to be ignored, but for accurate measurements of retention volume, and particularly capacity ratios, the actual volume measured should always be corrected for the extra column volume of the system and equation (11) should be put in the form

$$V_r = V_m + KV_S + V_E \qquad (13)$$

The exact nature of the dead volume is complex and, in fact, will vary from solute to solute due to the exclusion properties of the stationary phase, particularly if the stationary phase or support is silica or silica based. Thus, to measure (V_o) accurately, a non-adsorbed solute of the *same molecular size* as the solute should be used and then the correct retention volume (V'_r) can be calculated and employed for identification purposes.

However, providing

$$V'_r >>> V_o$$

then any non-adsorbed solute can be used to measure (V_o), and

$$V'_r = V_r - V_o = KV_S \quad (14)$$

Thus, (V'_r) is directly proportional to (K) and as (K) will be a unique property of a given solute, then (V'_r) can be used for solute identification purposes.

For example for two solutes (A) and (B) their corrected retention volumes will be will be

$$V'_{r(A)} = K_{(A)}V_S \quad (15)$$

and

$$V'_{r(B)} = K_{(B)}V_S \quad (16)$$

and both $V'_{r(A)}$ and $V'_{r(B)}$ can be used to identify solutes (A) and (B).

The Capacity Ratio of a Solute

The capacity ratio of a solute (k') was introduced early in the development of chromatography theory and was defined as the ratio of the distribution coefficient of the solute to the phase ratio (a) of the column. In turn the phase ratio of the column was defined as the ratio of the volume of mobile phase in the column to the volume of stationary phase in the column.

Thus, $$k' = \frac{V'_r}{a}$$

and, as $$a = \frac{V_s}{V_m}, \quad \text{then} \quad k' = \frac{KV_s}{V_m}$$

Note that (V_m) is the volume of mobile phase in the column and not V_o the total dead volume of the column.

Consequently, in practice

$$k' = \frac{V'_r}{V_o - V_E} \tag{17}$$

Equation (17) illustrates the importance of knowing the extra column volume to obtain accurate values for (k'). Nevertheless it must be pointed out that, in calculating (k'), the value taken in practice is often the ratio of the corrected retention distance (the distance in centimeters on the chart, between the dead point and the peak maximum) to the dead volume distance (the distance in centimeters on the chart, between the injection point to the dead point on the chromatogram). This calculation assumes the extra column dead volume is not significant and, as stated above, under certain conditions this may be true. Where computer data processing is used and no chart is available, the distances defined above would be replaced by the corresponding times in the software algorithm. The introduction of the capacity ratio for solute identification is an improvement over corrected retention volume as it is a dimensionless constant for any given column. However, the capacity ratio still depends on the *phase ratio* which will vary from column to column. As a consequence, capacity ratio data from one column *cannot* be used with confidence to identify a solute chromatographed on a different column.

The Separation Ratio

An alternative measurement, the *separation ratio* (α), was proposed to eliminate the effect of different phase ratios and different flow rates that could occur between different columns. The separation ratio is the ratio of the corrected retention volumes of two solutes that is the ratio of the respective retention volumes minus the dead volume. The measurement can be made directly on the chromatogram or, if the computer data processing is employed, the software will calculate it from the appropriate retention times.

Thus, for two solutes (A) and (B),

$$\alpha = \frac{V'_{r(A)}}{V'_{r(B)}} = \frac{K_{(A)}V_{s(A)}}{K_{(B)}V_{s(B)}}$$

where $V_{S(A)}$ is the volume of stationary phase available to solute (A) and $V_{S(B)}$ is the volume of stationary phase available to solute (A)

If exclusion effects are ignored or the solutes (A) and (B) have approximately the same molecular size then

$$V_{S(A)} = V_{S(B)}$$

and

$$\alpha = \frac{K_{(A)}}{K_{(B)}}$$

It is seen that the separation ratio is *independent* of the *phase ratios* of the two columns and the *flow rates* employed. It follows that the separation ratio of a solute can be used more reliably as a means of solute identification. Again, if the data is being processed by a computer, the corrected retention times will be used to calculate the separation ratios. In practice, a standard substance is often added to a mixture and the separation ratio of the substance of interest to the standard is used for identification.

However, it must be emphasized that retention data, whether they be corrected retention volumes, capacity ratios or separation ratios, do *not* provide unambiguous solute identification. Matching retention data between a solute and a standard obtained from two columns employing different phase systems would be more significant. Even so, it is still possible to have co-eluting solutes of different types on two different phase systems. Unambiguous identification would require supporting evidence, evinced by mass spectra, IR spectra or NMR spectra and such support would be essential for forensic purposes. In many instances the homogeneity of the peak must also be confirmed by taking appropriate spectra across the whole solute band during elution, e.g., by using a *diode array detector*. The use of the diode array detector for this purpose will be discussed later.

Column Efficiency

It is now necessary to attend to the second important function of the column. It has already been stated that, in order to achieve the separation of two substances during their passage through a chromatographic column, the two solute bands must be moved apart and, at the same time, must be kept sufficiently narrow so that they are eluted discretely. It follows, that the extent to which a column can constrain the peaks from spreading will give a measure of its quality. It is, therefore, desirable to be able to measure the peak width and obtain from it, some value that can describe the column performance. Because the peak will be close to Gaussian in form, the peak width at the points of inflexion of the curve (which corresponds to twice the standard deviation of the curve) will be determined. At the points of inflexion

$$\frac{d_2\left(X_0 \frac{e^{-v} v^n}{n!}\right)}{dv^2} = 0$$

or

$$\frac{d_2\left(X_0 \frac{e^{-v} v^n}{n!}\right)}{dv^2} =$$

$$X_0 \frac{e^{-v} v^n - e^{-v} n v^{(n-1)} - e^{-v} n v^{(n-1)} + e^{-v} n(n-1) v^{(n-2)}}{n!}$$

Simplifying and factoring the expression

$$\frac{d^2\left(X_0 \frac{e^{-v} v^n}{n!}\right)}{dv^2} = X_0 \frac{e^{-v} v^{(n-2)}(v^2 - 2v + n(n-1))}{n!}$$

Thus, at the points of inflexion,

$$v^2 - 2nv + n(n-1) = 0$$

and

$$v = \frac{2n \pm \sqrt{(4n^2 - 4n(n-1))}}{2}$$

$$= \frac{2n \pm \sqrt{4n}}{2}$$

$$= n \pm \sqrt{n}$$

It is seen that the points of inflexion occur after $n - \sqrt{n}$ and $n + \sqrt{n}$ plate volumes of mobile phase has passed through the column. Thus the volume of mobile phase that has passed through the column *between* the inflexion points will be

$$n + \sqrt{n} - n + \sqrt{n} = 2\sqrt{n} \tag{18}$$

Thus, the peak width *in milliliters* of mobile phase will be obtained by multiplying by the *plate volume* i.e.,

$$\text{Peak Width} = 2\sqrt{n}\,(v_m + K\,v_s) \tag{19}$$

The peak width at the points of inflexion of the elution curve is twice the standard deviation and thus, from equation (18) it is seen that the variance (the square of the standard deviation) is equal to (n), the total number of plates in the column. Consequently, the variance of the band (σ^2) in milliliters of mobile phase is given by

$$\sigma^2 = n(v_m + Kv_s)^2$$

Now, $$V_r = n(v_{m_s} + Kv_s)$$

Thus, $$\sigma^2 = \frac{V_r^2}{n}$$

It follows that as the variance of the peak is inversely proportional to the number of theoretical plates in the column then the larger the number of theoretical plates, the more narrow the peak and the more efficiently the column has constrained the band dispersion. As a consequence the number of theoretical plates in a column has been given the term *Column Efficiency* and is used to describe its quality.

Figure 10

A Chromatogram Showing the Separation of Two Solutes

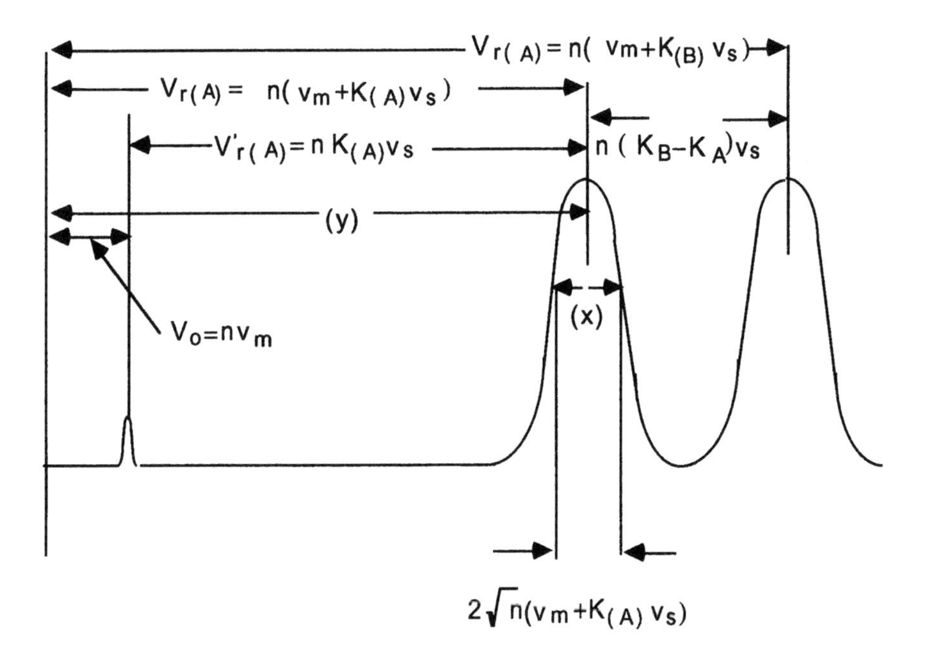

The various characteristics of the chromatogram that have so far been considered are shown in figure 10. It is important to be able to measure the efficiency of any column and, as can be seen from figure 10, this can be carried out in a very simple manner. Let the distance between the injection point and the peak maximum (the retention distance on the chromatogram) be (y) cm and the peak width at the points of inflexion be (x) cm as shown in figure 10.

Then as the retention volume is $n(v_m + Kv_s)$ and twice the peak standard deviation at the points of inflexion is $2\sqrt{n}(v_m + Kv_s)$, then, by simple proportion,

$$\frac{\text{Ret. Distance}}{\text{PeakWidth}} = \frac{y}{x} = \frac{n\,(v_m + Kv_s)}{2\sqrt{n}\,(v_m + Kv_s)} = \frac{\sqrt{n}}{2}$$

$$n = 4\left(\frac{y}{x}\right)^2 \qquad (20)$$

Equation (20) allows the efficiency of any solute peak, from any column, to be calculated from measurements taken directly from the chromatogram. Many peaks, if measured manually, will be only a few millimeters wide and, as the calculation of the column efficiency requires the width to be squared, the distance (x) must be determined very accurately. The width should be measured with a comparitor reading to an accuracy of ± 0.1 mm.

Figure 11 Measurement of Peak Width

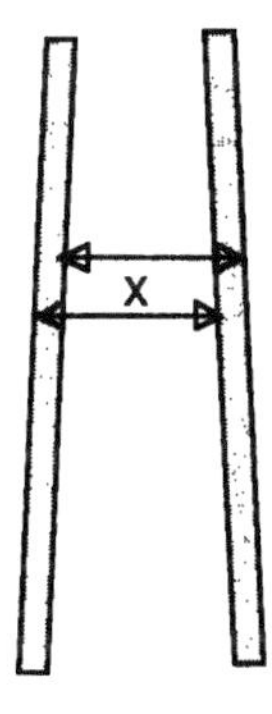

The measurements are taken from the inside of one line to the outside of the adjacent line, in order to eliminate errors resulting from the finite width of the ink line drawn on the chart. This procedure is shown in figure 11. The measurement should be repeated using the alternate edges of the line and an average taken of the two readings to avoid errors arising from any variation in line thickness. At least three replicate runs should be made and the three replicate values of efficiency should not differ by more than 3%. If the data acquisition system has software to measure efficiency then this can be used providing its accuracy is carefully checked manually. Noise on the detector can often introduce inaccuracies that are less likely to occur with manual measurement.

It is clear that the first challenge facing the analyst is the choice of the phase system that is appropriate for the particular sample to be analyzed. Only after the phase system has been chosen, can the correct column be selected. It is therefore necessary to know the types and properties of the different stationary phases that are available and how to formulate the pertinent mobile phases that must be used with them.

References

1/"*Liquid Chromatography Column Theory*", R.P.W.Scott, John Wiley and Sons, Chichester-New York-Brisbane-Toronto-Singapore, (1992).

2/ A.J.P.Martin and R.L.Synge, *Biochem. J.,* **35**(1941) 1358.

3/"*Gas Chromatography " Second Edition,* A.I.M. Keulemans (Ed. C.G.Verver) Reinhold Publishing Corporation, NewYork, (1959)106.

4/"*Theory and Mathematics of Chromatography*", A.S.Said, Dr. Alfred Hüthig Veriag GmbdH Heidelberg (1981)126.

5/ "*Ion Chromatography*", H. Small, Plenum Press (1989).

6/ *"Handbook of Ion Chromatography"*, J.Weiss, Published by Dionex, (1988).

7/*"Ion Chromatography Applications"*, R. E. Smith, CRC Press, (1988).

8/ R.P.W. Scott and P.Kucera, *J. Chromatogr.*, **125**(1976) 251.

9/ R. P. W. Scott and P. Kucera, *J. Chromatogr.*, **169**(1979) 51.

10/C.Peterson and G.Schill, *J Chromatogr.*,**204**(1981) 179.

11/ C.Peterson, *J. Chromatogr.***237**(1984) 553.

12/ D.R.Taylor, *J. Chromatogr. Sci.***30**(1992) 67.

13/ *"Cyclobond handBook"*,T.E.Beesley, Published by ASTEC (1991).

3
Liquid Chromatography Phase Systems

It was explained in the previous chapter that solute retention and, consequently, solute selectivity is accomplished in an LC column by exploiting three basic and different types of molecular interactions in the stationary phase; those interactions were described as ionic, polar and dispersive.

At this point it is necessary to discuss some alternative terms for molecular forces and molecular interactions that have crept into chromatography and that are alternatives to the presently accepted nomenclature. The reason for their introduction is unclear, but they appear to have become established in the biotechnology field. It is possible that, in contrast to biotechnology, which has evolved from organic chemistry and/or biochemistry, chromatography has evolved from a mixture of physical chemistry and analytical chemistry and, as a consequence, adopted an entirely different nomenclature. In any event, the alternative terms do cause confusion to those who are new to the techniques and are trying to learn more about them. In particular, they are confusing to those chemists that have had classical training and have been introduced to the correct and established physical chemical terms for describing molecular forces and molecular interactions.

The four terms that need discussion are (1) "hydrophobic interactions", (2)"lyophobic interactions", (3)"hydrophilic interactions" and (4) "lyophilic interactions".

The term "hydrophobic interaction" is a most unfortunate term in that it implies some form of molecular repulsion which of course, outside the Van der Waals radii of a molecule, is impossible. The term "hydrophobic force", literally meaning "fear of water" force, was coined, for some reason, as an alternative to the well established term, *dispersive* force. This may have been provoked by the immiscibility of a dispersive solvent such as n-heptane, with a very polar solvent such as water.

The reason that n-heptane and water are immiscible is *not* because water molecules *repel* heptane molecules. They are immiscible because the forces between two heptane molecules and the forces between two water molecules are much greater than the forces between a heptane molecules and a water molecule. Thus, water molecules and heptane molecules interact *very much more strongly* with *themselves* than with *each other*. One would have thought that the fact that water has a small but finite solubility in n-heptane, and n-heptane has a small but finite solubility in water, would have precluded repulsion as an interpretation.

Thus,

hydrophobic interactions are *dispersive interactions*

and *hydrophobic* solvents and stationary phases are *dispersive* solvents and stationary phases.

The term "hydrophilic force", literally meaning "love of water" force, appears to have been introduced merely as the complement to "hydrophobic". It is equivalent to the term polar, and polar solvents are hydrophilic solvents because they interact strongly with water or other polar solvents.

Therefore,

hydrophilic interactions are *polar interactions*

and *hydrophilic* solvents and stationary phases are *polar* solvents and stationary phases.

The reasons for the introduction of the terms "lyophobic" (meaning fear of lye) and "lyophilic" (meaning love of lye) are even more obscure and appear irrelevant as they are virtually alternatives to the terms hydrophobic and hydrophilic. The terms originated in the early soap industry during the mid-to-late nineteenth century. In about 1850 soap was prepared by boiling a vegetable oil with an alkaline solution obtained from leaching 'wood ash' with water.

The alkaline product from the wood ash was a crude solution of sodium and potassium carbonates called "lye". On boiling the vegetable oil with the lye, the soap (sodium and potassium salts of long chained fatty acids) separated from the lye due to the dispersive interactions between the of the fatty acid alkane chains and were thus, called "lyophobic". It follows that "lyophobic", from a physical chemical point of view, would be the same as "hydrophobic", and interactions between hydrophobic and lyophobic materials are dominantly dispersive. The other product of the soap making industry was glycerol which remained in the lye and was consequently, termed "lyophilic". Thus, glycerol mixes with water because of its many hydroxyl groups and is very polar and hence a "hydrophilic" or "lyophilic" substance.

Therefore,

lyophilic interactions are *polar interactions*

and

lyophobic interactions are *dispersive interactions.*

Perhaps we should consider it fortunate that, to date, exclusion chromatography has not yet been dubbed

"claustrophobic interaction chromatography".

Finally, before returning to the subject of the chapter, the extravagant and apparently uncontrollable contemporary use of acronyms must be mentioned. The reason for their excessive use is not clear but, for whatever reason, they can be extremely frustrating and annoying impediments to the student or scientist who is studying a new subject. Their excessive use is to be deplored and in the following discussion and, in fact, throughout this book, acronyms will be kept to an absolute minimum and where used, will be clearly and unambiguously defined.

Returning now to the subject of the chapter, in addition to appropriate retentive characteristics, a potential stationary phase must have other key physical characteristics before it can be considered suitable for use in LC. It is extremely important that the stationary phase is completely insoluble (or virtually so) in all solvents that are likely to be used as a mobile phase. Furthermore, it must be insensitive to changes in pH and be capable of assuming the range of interactive characteristics that are necessary for the retention of all types of solutes. In addition, the material must be available as solid particles a few microns in diameter, so that it can be packed into a column and at the same time be mechanically strong enough to sustain bed pressures of 6,000 p.s.i. or more. It is clear that the need for versatile interactive characteristics, virtually universal solvent insolubility together with other critical physical characteristics severely restricts the choice of materials suitable for LC stationary phases.

To date there have been three possible candidates that will generally fit the requirements of an LC stationary phase and they are alumina, silica gel and certain polymeric resins for example, cross-linked polystyrene. Alumina, despite having an abundance of surface hydroxyl groups that confer on it strong polar characteristics and despite it having adequate mechanical strength, is the least useful of the three materials for LC. Although alumina can be used as a stationary phase *per se,* it does not readily form stable bonded phases though it can be coated with polymeric stationary phases which endow it with some dispersive character. Furthermore, although it can be ground to particle sizes of a few microns in diameter, a procedure has not been found that can produce the high column efficiencies that are consistently attainable from silica gel. As a result alumina stationary phases are still utilized in only a few per cent of all LC applications.

The polymeric resin beads fill a need that arises from the instability of silica gel and its products to mobile phases of extreme pH (outside a pH range of about 4.0-7.0) and, consequently, are employed in most ion exchange separations. Organic moieties containing ionic groups can be bonded to silica and produce an effective ion exchange media, but the restrictions of pH on phase stability still apply. It follows that ion exchange bonded phases are less popular than the polymer bead alternatives.

Nevertheless, silica gel is the material of choice for the production of the vast majority of LC stationary phases. Due to the reactive character of the hydroxyl groups on the surface of silica gel, various organic groups can be bonded to the surface using standard silicon chemistry. Consequently, the silica gel surface can be modified to encompass the complete range of interactive properties necessary for LC ranging from the highly polar to almost completely dispersive.

The Production of Silica Gel

As a result of its unique chemical and physical properties, silica gel is probably the most important single substance involved in liquid chromatography today. Without silica gel, it is doubtful whether HPLC could have evolved at all. Silica gel is an amorphous, highly porous, partially hydrated form of *silica* which is a substance made from the two most abundant elements in the earth's crust, *silicon* and *oxygen*. Silica, from which silica gel is manufactured, occurs naturally, either in conjunction with metal oxides in the form of silicates, such as clay or shale, or as free silica in the form of quartz, cristobalite or tridymite crystals. Quartz is sometimes found clear and colorless, but more often in an opaque form, frequently colored yellow by traces of iron which, when crushed by earth movement and weathered by air and water, produces sand.

Silica gel is prepared in two stages. Firstly, an intimate mixture of sand and sodium carbonate are heated together in iron pans forming sodium silicate with the release of carbon dioxide. The sodium silicate is leached out of the cooled mixture, unreacted sand removed by filtration, and the resulting silicate solution treated with hydrochloric or sulfuric acid.

Initially, silicic acid is released,

$$Na_2SiO_3 + H_2O + 2HCl = Si(OH)_4 + 2NaCl$$

However, the free acid quickly starts to condense with itself, accompanied by the elimination of water to form dimers, trimers and eventually polymeric silicic acid. The polymer continues to grow, initially forming polymer aggregates and then polymer spheres, a few Angstroms in diameter. These polymeric spheres are termed the *primary* particles of silica gel and must not to be confused with the macro-particles of silica gel that are packed into the LC column.

The primary particles continue to develop until, at a particular size, the surface silanol groups (hydroxyl groups attached to the surface silicon atoms of the primary particles) on adjacent primary polymer particles, condense with the elimination of water causing the primary particles to fuse together and the solution begins to gel. During this process, the primary particles of silica gel that are formed will have a range of diameters extending from a few Angstroms to many thousands of Angstroms depending on the conditions of formation. It is the size of the primary particles that determines the pore size and pore range of the final gel. The gel formed in the manner described above is termed the hydrogel and at this stage is very soft with little or no mechanical strength.

The hydrogel is allowed to stand for a few days during which time a process called *sinerisis* takes place. During sinerisis the condensation of the primary particles, one with another, continues and the gel shrinks further, accompanied by the elimination of more saline solution that exudes from the gel. After three or four days, sinerisis is complete and the gel becomes firm and can now be washed free of residual electrolytes with water. The washed product is finally heated to 120°C to complete the condensation of the surface silanol groups between the particles, and a *hard* xerogel is formed. It is this xerogel that is used as the LC stationary phase and for bonded phase synthesis. It is not intended to discuss the production of silica gel in detail and those interested are referred to "Silica Gel and Bonded Phases", published by Wiley (1).

However, it should be said in passing that the pore size and surface area of the silica, which can be critical for certain LC applications, is controlled by the conditions of gelling, the subsequent washing conditions and any ensuing thermal treatment.

Silica gel, in the form of the xerogel, is an extremely hard and abrasive material and thus, must be ground by an appropriate mill. For effective use as an LC stationary phase the so called "irregular" particles produced by the mill should be well rounded to simulate spherical particles. In fact, spherical particles of silica gel can be produced directly (2-4), but the procedure is tedious and more expensive. Rounded, irregular silica gel particles, will provide column efficiencies at least as good, if not better, than those obtained from the spherical material. The ideal mill for grinding the xerogel for LC purposes is the "Jet Mill" which can provide narrow cuts of particles having mean diameters that range from 2 to 20μ. The size reduction is achieved by entraining the particles of grist into three (or if necessary more) jets of air that are directed to a center point in a cylindrical mill housing. The particles collide at the center, causing self fragmentation with minimum wear on the grinding system and very little product contamination. Due to the nature of the grinding process it is particularly useful for grinding very hard and abrasive materials.

The product from the "Jet Mill" after classification can be used directly as an LC packing or as the raw material for bonded phase synthesis.

The silica particles are packed into the LC column by a slurry packing procedure using either a 'balanced density' solvent or a high viscosity solvent. The principal of the packing procedure is to complete the packing before the particles are segregated by sedimentation. Thus, a solvent mixture (usually a mixture of halogenated paraffins) having the same density as the silica gel particles can be used so that no sedimentation takes place at all. Alternatively, a viscous solvent can be employed to ensure that the sedimentation is so slow that the packing is complete before any significant size segregation has taken place. In most cases, analysts will purchase *packed* columns and thus, details of packing procedures are not pertinent to this discussion; those interested in slurry packing methods are again referred to "Silica Gel and Bonded Phases", published by Wiley (1).

Solute/Solvent Interactions on the Surface of Silica Gel

Mono-Layer Adsorption of Solvents on the Surface of Silica Gel

As a result of the hydroxyl groups that cover its surface, silica gel is strongly polar and thus, would be useful for separating polarizable or weakly polar solutes. In a practical separation, to ensure that polar selectivity *dominates* in the *stationary phase* and polar interactions in the mobile phase are minimized, the mobile phase must be chosen to be non-polar and strongly dispersive, for example n-heptane.

However, consider the separation of solutes that are more polar than the aromatic hydrocarbons, for example mixtures of ethers or aliphatic esters. If it were attempted to separate a series of aliphatic esters on silica gel employing n-heptane as the mobile phase it would be found that the retention of the later eluting solutes was inordinately long. If a slightly stronger solvent, such as chloroform was used as an alternative to n-heptane, it would be found that the less polar esters were eluted too rapidly. Consequently, a suitable mixture of chloroform and n-heptane might be considered more appropriate. Although a mixture of chloroform and n-heptane would, indeed, be very suitable for the separation of the esters, the effect of the addition of chloroform to the n-heptane on solute retention is by no means simple. When chloroform is added to n-heptane, it is found that not only is the eluting capability of the mobile phase mixture increased due to the stronger interaction of the solutes in the mobile phase with chloroform, but the nature of the silica surface is also changed and the retentive capacity of the stationary phase is reduced. This is because the *solvents* themselves partition between the silica surface and the solvent mixture as well as the *solutes*. In fact as the chloroform concentration increases, some is adsorbed on the silica surface as a monolayer according to the Langmuir adsorption isotherm.

The equation for the Langmuir adsorption isotherm is as follows:

$$\alpha = \frac{ac}{b+ac} \tag{1}$$

where (α) is the fraction of the surface covered by the solute,
(c) is the concentration of the solute in the mobile phase,
and (a) and (b) are constants.

It is seen that at high concentrations (α) becomes unity and the surface is completely covered with the more strongly adsorbed solvent. The adsorption isotherm of chloroform on silica gel, determined by Scott and Kucera (5) is shown in figure 1. It is seen that the monolayer of chloroform collects on the surface continuously until the chloroform content of the mobile phase is about 50%. At this concentration the monolayer appears complete. Thus, between 0 and 50% chloroform in the n-heptane, the interactions between the solute and the chloroform in the mobile phase are continuously increasing.

Figure 1

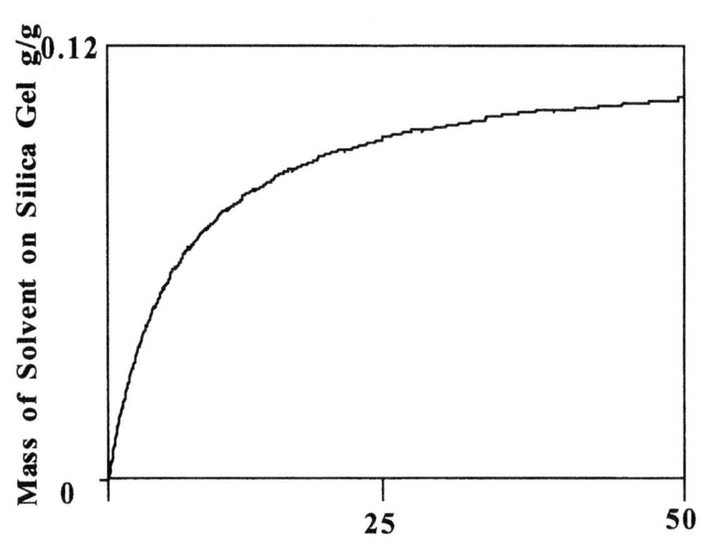

In contrast, the interactions with the stationary phase are becoming weaker as the surface becomes covered with chloroform. Thus retention is reduced by both the increased interactions in the mobile phase and reduced interaction with the stationary phase. When the concentration of chloroform in the solvent mixture is in excess of 50%, then the interactive properties of the stationary phase no longer change as the surface is now covered with a mono-layer of chloroform. However, solute retention will continue to decrease due to the increased interactions of the solute with the higher concentrations of chloroform in the mobile phase. It is clear that even with this simple example the dependence of retention on solvent composition is quite complex.

Solute Interactions with the Silica Gel Surface
(Mobile Phase n-Heptane/Chloroform)

There are basically two types of interaction that can take place between a solute and the silica gel surface. Firstly, the solute molecule can interact with the adsorbed solvent layer and rest on the surface of it. This type of interaction is called *sorption interaction* and occurs when the molecular forces between the solute and the silica are relatively weak compared with the forces between the solvent molecules and the silica. The second type of interaction is where the solute molecules displace the solvent molecules from the surface and interact directly with the silica gel itself, for example, the silanol groups. This type of interaction is called *displacement interaction* and occurs when the interactive forces between the solute molecules and the silica surface are much stronger than those between the solvent molecules and the silica surface.

The types of interaction that can occur between the solute and the stationary phase surface when in contact with the pure solvents, n-heptane, chloroform or a mixture of n-heptane/chloroform are shown in figure 2.

In pure n-heptane or pure chloroform the solute molecules can either interact directly with the surface of the adsorbed solvent or displace the adsorbed solvent and interact directly with the silica surface. In the case of the solvent mixture the solute molecules may interact with the surface of either solvent or displace either solvent and interact directly with the silica surface or any combination of these possibilities. For example some solute molecules might displace the layer of n-heptane and interact directly with the surface. At the same time, those solute molecules striking the layer of chloroform may interact only with the chloroform and not be capable of displacing it, as the molecular forces between the chloroform and the silica gel are greater than the molecular forces between the solute and the silica gel.

From a practical point of view the change in retention with solvent concentration will be greater at the lower concentrations of chloroform where interactions in both phases are being changed. At concentrations above 50%, however, interactions are only changing in the mobile phase and so the effect of solvent concentration on retention will be less significant.

Figure 2

Solute/Solvent Interactions with a Silica Gel Surface

Mobile Phase n-Heptane

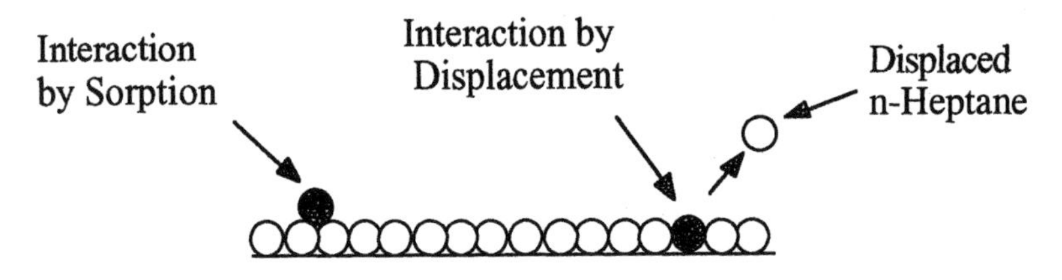

Mobile Phase
n-Heptane/Chloroform

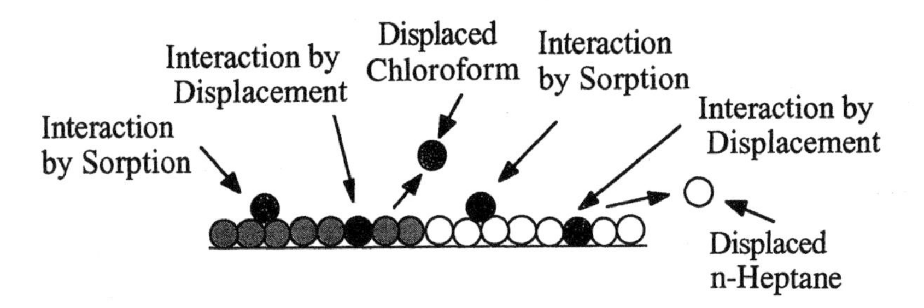

Mobile Phase Chloroform

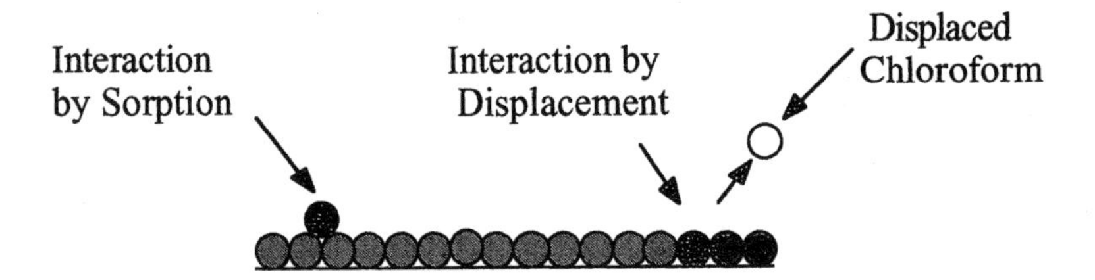

This will be true for all solvent mixtures where there is no association between solvents and is useful to remember when adjusting solvent composition to obtain the desired selectivity and retention. Unfortunately, when one solvent is polar, a bi-layer of solvent can be formed on the surface and solute/solvent interactions with the silica gel are even more complex.

Bi-Layer Adsorption of Solvents on the Surface of Silica Gel

The identification of bi-layer adsorption of polar solvents on the surface of silica gel arose from some work by Scott and Kucera (5) who measured the adsorption isotherms of the some polar solvents, ethyl acetate, isopropanol and tetrahydrofuran from n-heptane solutions onto silica gel. The authors found that the experimental results for the more polar solvents did *not* fit the simple mono-layer adsorption equation and, as a consequence, the possibility of bi-layer adsorption on the silica gel surface was examined.

Bi-layer adsorption is not uncommon and the development of the bi-layer adsorption isotherm equation is a simple extension of that used for the mono-layer equation. The Langmuir equation for bi-layer adsorption is as follows:

The equation for the *first layer,*

$$\alpha = \frac{bdc + adc^2}{eb + bdc + adc2} \qquad (2)$$

where (α) is the surface fraction covered by the first adsorbed layer,
(c) is the concentration of the polar solvent in n-heptane
and (a),(b) and (d) are constants.

The equation for the *second layer,*

$$\beta = \frac{ac}{b + ac} \qquad (3)$$

where (β) is the surface fraction covered by the second adsorbed layer, and the other symbols have the meaning previously ascribed them.

Figure 3

The Individual and Combined Adsorption Isotherms for Ethyl Acetate on Silica Gel

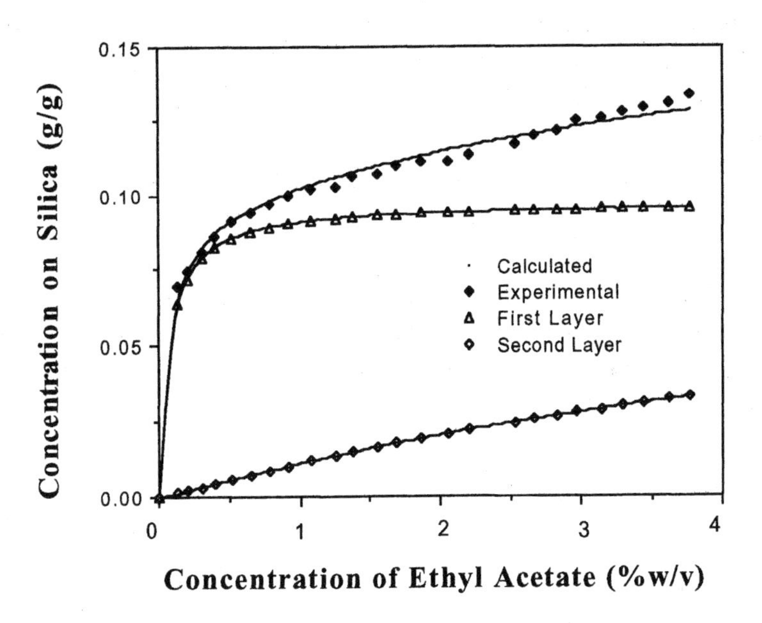

Scott and Kucera fitted their data to equations (2) and (3), solved for the constants and reconstructed the isotherms for each layer which are included in figure 3. The formation of a solvent bi-layer is diagramatically represented in figure 4.

Figure 4

Bi-layer Adsorption of Solvent

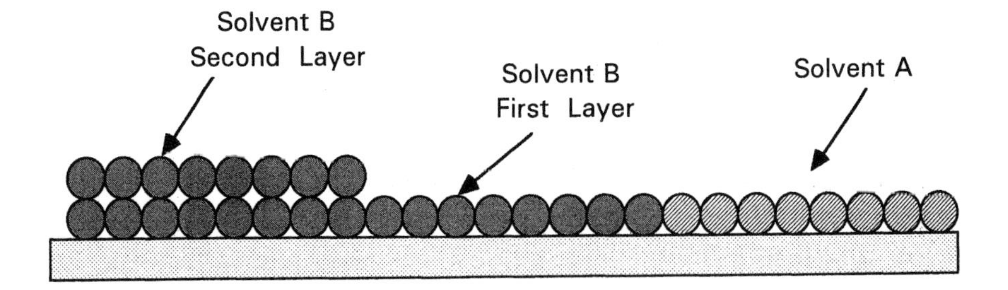

It is seen in figure 3 that the two curves, although of the same form, are quite different in magnitude. The first layer to be formed is very strongly held to the surface and is almost complete when the concentration of ethyl acetate in the mobile phase is only about 1%w/v. As the concentration of ethyl acetate rises above 1%w/v the second layer is only just being formed. The formation of the second layer of ethyl acetate is much slower and obviously the interactions between the solvent molecules with those already adsorbed on the surface are much weaker than their interaction with the silanol groups of the silica gel.

This means, in practice, that when employing a polar solvent with n-heptane (or any other paraffin for that matter) to reduce the retention, there will be a dramatic reduction in retention over the concentration range of about 0-2%w/v. However, subsequent changes in solute retention with polar solvent concentration will be relatively small. This will be true for any polar solute and was experimentally verified by Scott and Kucera for solutions of ethyl acetate, tetrahydrofuran and n-propanol in n-heptane. The very sensitive relationship between solvent concentration and retention at very low concentrations makes the phase system very difficult to make reproducible. This problem is one of the factors that deter analysts from using silica gel as a stationary phase for the separation of polar solutes. It is very satisfactory, however, for the separation of polarizable and weakly polar substances that can be eluted by paraffin/methylene dichloride or similar types of solvent mixtures.

Solute Interactions with the Silica Gel Surface
(Mobile Phase n-Heptane/Ethyl Acetate)

The possible alternatives for a solute molecule to interact with a bi-layer of solvent molecules is depicted in figure 5. It is seen that such a surface offers a wide range of sorption and displacement processes that can take place between the solute and the stationary phase surface. There are, in fact, three different surfaces on which a molecule can interact by sorption and three different surfaces from which molecules of solvent can be displaced and allow the solute molecule to penetrate.

Figure 5

Different Types of Solute Interaction that Can Occur on Silica Surfaces Covered with a Solvent Bi-layer

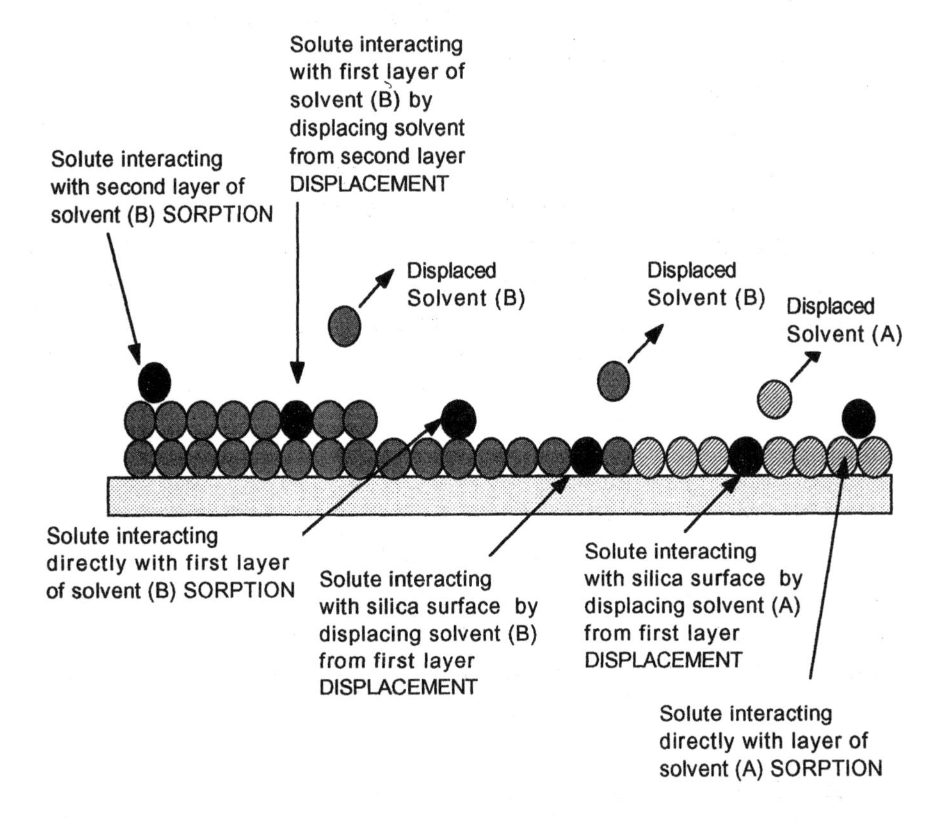

In any separation all the alternatives are possible, but it is more likely that for any particular solute, one type of interaction will dominate. Where there are multi-layers of solvent, the most polar is the solvent that interacts directly with the silica surface, and consequently constitutes the first layer. Depending on the concentration of the polar solvent, the next layer may be a second layer of the same polar solvent as in the case of ethyl acetate. If, however, the quantity of polar solvent is limited, then the second layer might consist of a less polar component of the solvent mixture. If a ternary mixture of solvents is used, the nature of the surface, and the solute interactions with the surface can become very complex indeed. In general the stronger the polarity of the solute the more likely it is to interact with the surface by displacement even to the extent of displacing both layers of solvent (one of the alternative processes that is not depicted in figure 5). Less polar solutes a more likely to interact with the surface by sorption.

Silica Gel as an Exclusion Medium

Silica gel can be produced with a wide range of pore sizes and can thus be used as an exclusion media as discussed in chapter 2. An example of the different exclusion properties exhibited by silica gel is shown in figure 6.

Figure 6

Graph of Pore Volume per Gram of Silica Gel against Log Pore Diameter for Ten Different Silica Gels

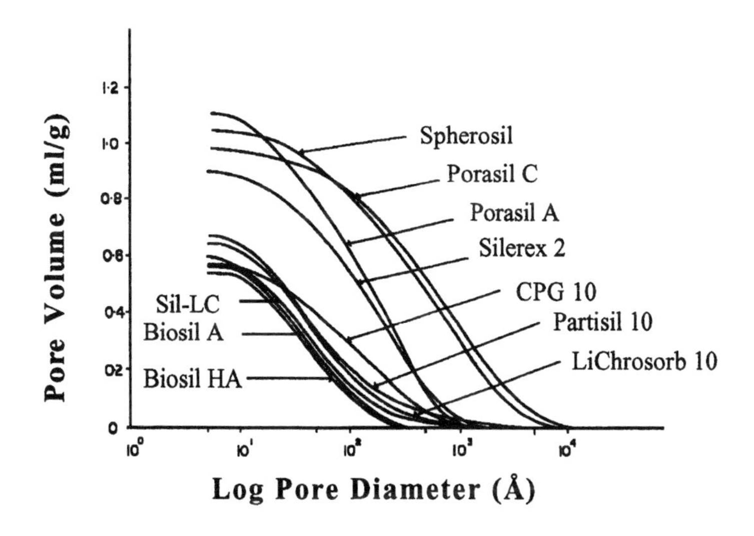

It is seen that the pore diameter of the silica gels cover a range from about 7 to 250 Angstroms typified by Partisil to that typified by Porasil C ranging from about 50 to over 1,000 Angstroms. It would seem that silica would be a valuable media for exclusion chromatography. Unfortunately, the main application for exclusion chromatography lies in the biotechnology field in the separation of macro-molecules, where the mobile phases that can be employed must be aqueous in nature to permit adequate solubility and solute stability. Silica gel is soluble in water, albeit very slightly, which renders a column packed with such material unstable. Furthermore, the silica gel surface is very active which would be liable to denature large peptides and proteins during the separation procedure. Nevertheless, silica gel is used for exclusion chromatography providing the materials are dispersive or weakly polar in nature and polar or semi-polar solvents can be employed. For example, silica gel can be used very elegantly to separate the larger aromatic hydrocarbons by exclusion using acetonitrile or tetrahydrofuran as the mobile phase or for the separation of polystyrene polymers using tetrahydrofuran as the mobile phase.

Figure 7

The Separation of a Mixture of Polyphenyl Siloxanes by Exclusion Chromatography

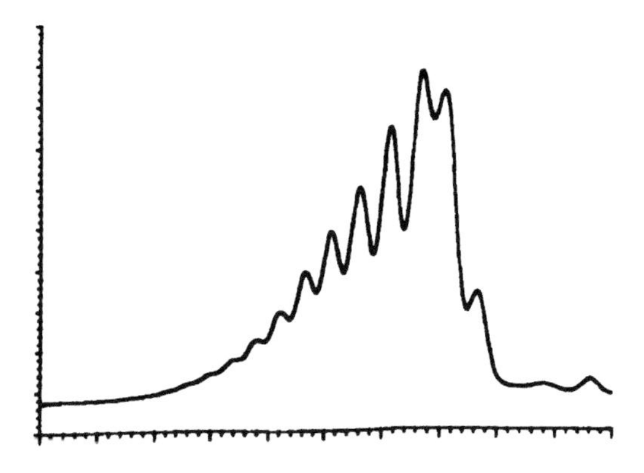

An example of the use of silica gel to separate the products from a phenyl siloxane polymerization employing a high efficiency column

(10m long, 1 mm i.d., packed with 20 m Partisil silica gel) is shown in figure 7. It is seen that the individual polymers are clearly separated but again a very a high efficiency was necessary and correspondingly long analysis times had to be tolerated due to the relatively small molecular weight intervals between each polymer. The chromatogram shown, covers a time period extending between 4 and 5 hours. However, it should be pointed out that silica gel can be used for the separation of dispersive and weakly polar solutes using standard columns involving separation times of *less* than an hour.

Application Areas for Silica Gel as a Stationary Phase

It is important for the analyst to be able to select the best stationary phase to use for a particular chromatographic analysis. Silica gel can be used in two modes of chromatographic separations: as a stationary phase in normal elution development or as a stationary phase in exclusion chromatography.

Silica Gel as a Stationary Phase in Elution Chromatography

As a result of its highly polar character, silica gel is particularly useful in the separation of polarizable materials such as the aromatic hydrocarbons and polynuclear aromatics. It is also useful in the separation of weakly polar solute mixtures such as ethers, esters and in some cases, ketones. The mobile phases that are commonly employed with silica gel are the n-paraffins and mixtures of the n-paraffins with methylene dichloride or chloroform. It should be borne in mind that chloroform is opaque to UV light at 254 nm and thus, if a fixed wavelength UV detector is being used, methylene dichloride might be a better choice. Furthermore, chloroform is considered toxic and requires special methods of waste disposal. Silica gel is strongly deactivated with water and thus, to ensure stable retentive characteristics, the solvent used for the mobile phase should either be completely dry or have a controlled amount of water present. The level of water in the solvent that will have significant effect on solute retention is extremely small. The solubility of water in n-heptane is only a few parts per million but, if present, will, in time, significantly deactivate silica gel.

If the mixture to be separated contains fairly polar materials, the silica may need to be deactivated by a more polar solvent such as ethyl acetate, propanol or even methanol. As already discussed, polar solutes are avidly adsorbed by silica gel and thus the optimum concentration is likely to be low, e.g. 1-4%v/v and consequently, a little difficult to control in a reproducible manner. Ethyl acetate is the most useful moderator as it is significantly less polar than propanol or methanol and thus, more controllable, but unfortunately adsorbs in the UV range and can only be used in the mobile phase at concentrations up to about 5%v/v. Above this concentration the mobile phase may be opaque to the detector and thus, the solutes will not be discernible against the background adsorption of the mobile phase. If a detector such as the refractive index detector is employed then there is no restriction on the concentration of the moderator. Propanol and methanol are transparent in the UV so their presence does not effect the performance of a UV detector. However, their polarity is much greater than that of ethyl acetate and thus, the adjustment of the optimum moderator concentration is more difficult and not easy to reproduce accurately. For more polar mixtures it is better to explore the possibility of a reverse phase (which will be discussed shortly) than attempt to utilize silica gel out of the range of solutes for which it is appropriate.

In summary, examples of the successful use of silica gel as a conventional stationary phase are in the analysis of mixtures containing polarizable and relatively low polarity solutes typified by mixtures of aromatic hydrocarbons, polynuclear aromatics, nitro compounds, carotenes and vitamin A formulas.

Silica Gel as a Stationary Phase in Exclusion Chromatography

In many respects silica gel is an ideal exclusion media. Due to its rigidity and mechanical strength, it will sustain high pressures and consequently, can be packed into long columns that provide the necessary high efficiencies. Furthermore, silica gel is available in a wide range of pore sizes which would make it suitable for the separation of substances having a wide range of molecular weights. Unfortunately it has two major disadvantages. Firstly it is slightly soluble in water and thus cannot be used for any length of time in aqueous solvent mixtures. Secondly it is very unstable at extremes of pH. As a consequence, it is precluded from the major areas of application for exclusion chromatography, that is the separation of macro-molecules of biological origin. Such samples require to be separated in aqueous solvents at pH values outside those to which silica gel are stable. Nevertheless, there are areas of application where silica gel is ideally suited as an exclusion media but, unfortunately, these areas are rather restricted. In order to ensure that the separation is by exclusion, and there are no differential interactions between the solutes and the two phases, a fairly polar solvent should be used to deactivate the silica gel. A typical solvent for use as the mobile phase when using silica gel as an exclusion medium is tetrahydrofuran.

In summary, silica gel can be an excellent stationary phase for use in exclusion chromatography in the separation of high molecular weight, weakly polar or polarizable polymers. It cannot be used for separating mixtures that require an aqueous mobile phase or operate at a pH outside the range of 4-8. Examples of the type of materials that can be separated by exclusion chromatography using silica gel are the polystyrenes, polynuclear aromatics, polysiloxanes and similar polymeric mixtures that are soluble and stable in solvents such as tetrahydrofuran.

Bonded Phases

Although silica gel was the catalyst that partly initiated the renaissance of liquid chromatography, it was quickly found that separations based solely on polarity had severe limitations. Furthermore, the labile nature of silica gel in aqueous environments coupled with its instability to extremes of pH provoked active research to develop alternative stationary phases. The immediate and obvious solution was to chemically modify the surface of silica gel by bonding organic moieties to it that would exhibit dispersive interactions and at the same time stabilize the material to water and extremes of pH. The first attempts to bond an organic group to silica gel for chromatographic purposes was made by Halasz and Sebastian (6), who attached aliphatic hydrocarbon chains to the surface by means of the silicon-oxygen-carbon linkage, a pioneering step in the development of liquid chromatography stationary phases. The synthesis by Halasz was carried out by refluxing the silica gel with a high molecular weight aliphatic alcohol, and thus, by esterifying the hydroxyl groups of the polymeric silicic acid, aliphatic silyl ester groups were attached to the silica gel surface. Unfortunately, the silicon-oxygen-carbon linkage is very weak and the bonded phase of Halasz rapidly hydrolyzed from the surface, thus, regenerating the original silica gel. Nevertheless, the attached aliphatic moiety was sufficiently stable to give Halasz the opportunity to identify the highly desirable chromatographic properties of such phases and his discovery can be ranked as one of the more important contributions to the evolution of LC. In the following year, Kirkland (7) described an alternative bonding process involving the use of chlorsilanes.

On reacting a chlorsilane with the hydroxyl groups of the silica gel the hydrocarbon chain is attached by the stronger silicon-oxygen-silicon link with the elimination of hydrogen chloride. An example of the reaction is given below. The radical (R) can be aliphatic, aromatic, heterocyclic or containing another organic groups but, unless protected, obviously no groups with which the chlorsilane would react.

$$-\overset{|}{\underset{|}{Si}}-O-H \quad + \quad R(CH_3)_2-Si-CL$$

$$\dashrightarrow \quad -\overset{|}{\underset{|}{Si}}-O-Si-R(CH_3)_2 \quad + HCl$$

In practice (R) is most frequently a straight aliphatic hydrocarbon chain 8 or 18 carbon atoms long that produces the well known *reverse phases*. For example, treating silica, dried at about 150°C, with dimethyloctylchlorsilane contained in a suitable solvent at elevated temperatures causes a dimethyloctylsilyl group to be attached to the surface by silicon-oxygen-silicon bonds and is one of the more popular reverse phases. This bonding sequence has become the standard for the production of all classes of bonded phases, although there are other methods of synthesis and there is now a wide range of silane reagents from which to choose. A set of reagents of particular interest are the alkoxyalkylsilanes which react in a similar manner to the chlorsilanes but, as one might expect, require somewhat different reaction conditions. The ethoxy- and methoxysilanes are the most reactive of the alkoxysilanes and, consequently, are the most commonly used for bonded phase synthesis.

The Different Classes of Bonded Phases

There are three classes of bonded phase, the specific class being determined by the type of silane reagent that is used in their preparation. The three types of silane reagents are those with a single reacting group, those with two reacting groups and those with three, e.g. the mono-, di- and trichlorsilanes.

The 'Brush' Phases

The 'brush' phases are made from the mono-chlorsilanes or the mono-alkoxysilanes which attach a single organic moiety to each reacted hydroxyl group. For example using octyldimethylchlorsilane, dimethyloctyl chains would be attached to the surface like bristles of a brush, hence the term 'brush' phase. In all bonding processes, some hydroxyl groups remain unreacted which would allow the polar character of the silica play a part in retention and also render the surface liable to dissolution in water and dilute acids. For this reason the material is 'capped' by treatment with either trimethylchlorsilane or hexamethyldisilazane to eliminate any unreacted or stearically hindered hydroxyl groups. An example of the surface of a "brush" phase is diagramatically represented in figure 8.

Figure 8

A 'Brush' Phase

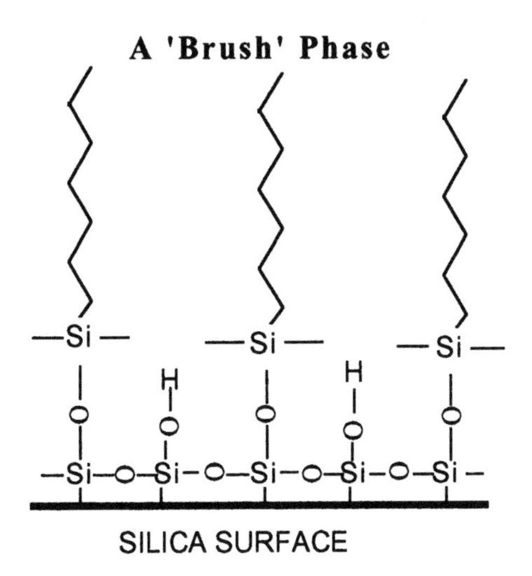

It is seen that each dimethyloctyl 'bristle' is represented as being separated from one another by an hydroxyl group. This concept represents 'contemporary thinking' as it is assumed that the hydroxyl groups are fairly regularly oriented on the silica surface and that the dimethyl chains stearically prevent adjacent hydroxyl group reacting with the chlorsilane, even when the reagent is in excess. Furthermore it is also assumed that the hydroxyl groups are stearically hindered from the capping reagents as well. This concept is only partially supported by experiment and remains somewhat open to question. In any event, this book is not the place to argue the point. The 'brush' type of phase appears to be the most popular, can be made in a fairly reproducible manner, is stable in aqueous solutions and pure water, but becomes unstable at a pH of less than 4 and in excess of 8.0

The Oligomeric Phases

If methyloctyldichlorsilane is used as the reagent in a sequence of synthetic steps, an oligomeric phase can be built up on the surface.

The silica is first reacted with the methyloctyldichlorsilane to link methyloctylchlorsilyl groups to the surface. The bonded phase is then treated with water to generate methyloctylhydroxysilyl groups which, in turn, are then reacted with more methyloctyldichlorsilane attaching another methyloctylchlorsilyl group to the previous group. This process can be repeated until eight or ten oligomers are linked to each other and attached to each stearically available hydroxyl group on the surface. The product is finally treated with trimethylchlorsilane or some other capping reagent to eliminate the last hydroxyl groups formed at the end of the oligomer. The oligomers are layered over the surface making the product extremely stable exhibiting almost no polar characteristics whatsoever. However, due to the complexity of the synthesis, oligomeric phases are expensive to manufacture and, consequently, are not often used and, at this time, are not commercially available. They may, however, become more accessible in the future as demand for bonded phases with good pH stability for the separation of materials of biological origin becomes more urgent.

The 'Bulk' Phases

If the silica surface is saturated with water and octyltrichlorsilane is used as the reagent, reaction occurs with both the hydroxyls of the silica surface and the adsorbed water, causing a cross-linking reaction and an octylsilyl polymer to be built up on the surface. The same procedure can be used as that in the synthesis of oligomeric phases and the material can be alternatively treated with water and the trichlorsilane reagent. Layers of bonded phase are built up on the surface but, in this case, due to the trichlor function of the reagent, cross-linking occurs. As a result of the polymerization process, the stationary phase has a chemically cross-linked, multi-layer character and, consequently, is termed a 'bulk' phase.

The "bulk" phases are almost as popular as the 'brush' phases as they tend to have a higher carbon content (more organic material bonded to the surface) and thus provide a little greater retention and selectivity. 'Bulk' phases have about the same stability to aqueous solvents and pH as the 'brush' phases.

Using appropriate organic chlorsilanes, polar or polarizable groups such as nitriles or aromatic rings can be bonded to the silica to provide stationary phases covering a wide range of polarities. Bonded ion exchange materials have also been synthesized, although they are not as stable to salt solutions and extremes of pH as the ion exchange resins.

An interesting carbon stationary phase, introduced by Knox (8), is obtained by filling the pores of appropriately sized silica particles with an organic polymer and carbonizing the product at elevated temperatures. The silica is removed from the product by treatment with strong alkali or hydrofluoric acid forming, perhaps, what might be termed a true 'reverse phase'. It would be a "reverse phase" in the sense that the pores are where the primary particles of silica existed and the solid matrix now replaces the pores. The product is too active for use in chromatography and so the carbon is graphitized by exposure to an argon plasma. This is a relatively new material and, due to the complexity of its manufacture, is expensive. Whether its performance relative to that of a conventional reverse phase merits the greater cost, at this time remains to be established.

Reverse Phases

The most popular bonded phases are, without doubt, the reverse phases which consist solely of aliphatic hydrocarbon chains bonded to the silica. Reverse phases interact dispersively with solvent and solute molecules and, as a consequence, are employed with very polar solvents or aqueous solvent mixtures such as methanol/water and acetonitrile/water mixtures. The most commonly used reverse phase appears to be the brush type phase with aliphatic chains having four, eight or eighteen carbon atom chains attached. These types of reverse phase have been termed C4, C8 and C18 phases respectively. The C8 and C18 phases are mainly used for separating solutes having relatively low molecular weights whereas the C4 phase is used for the separation of very large molecules.

The C4 reverse phase is also particularly useful in the separation of materials of biological origin that may be chemically labile or easily denatured. Due to the strong interactions that can take place between large polypeptide or protein molecules and the reverse phase, such compounds are often denatured or de-conformed after interaction with a reverse phase surface. In many instances, the de-conformation of the large macro-molecules is accompanied by biological deactivation and is often irreversible. It is clear that such deactivation must be avoided. The de-conformation appears to be related to the chain length of the reverse phase and for this reason, biotechnologists prefer to use the C4 reverse phases to minimize the possibility of any biologically active material being deactivated. More recently, C2 bonded phases have also been made available for this reason, which reduces the possibility of de-conformation still further. Unfortunately, the retentive capacity of the reversed phase is also reduced.

The Interaction of Reverse Phases with Solvents and Solutes

Solvents interact with reverse phases in very much the same way as they do with the surface of silica gel. However, in this case it is the more dispersive component of the mobile phase that is adsorbed on the surface as opposed to silica gel, which being a polar stationary phase, adsorbs the more polar solvent onto its surface.

The more dispersive solvent from an aqueous solvent mixture is adsorbed onto the surface of a reverse phase according to Langmuir equation and an example of the adsorption isotherms of the lower series of aliphatic alcohols onto the surface of a reverse phase (9) is shown in figure 9. It is seen that the alcohol with the longest chain, and thus the most dispersive in character, is avidly adsorbed onto the highly dispersive stationary phase, much like the polar ethyl acetate is adsorbed onto the highly polar surface of silica gel. It is also seen that the surface of the reverse phase is virtually covered when the concentration of butanol in the aqueous mixture is only about 4%w/v.

Figure 9

The Adsorption Isotherms of Some n-Alcohols on a Reverse Phase

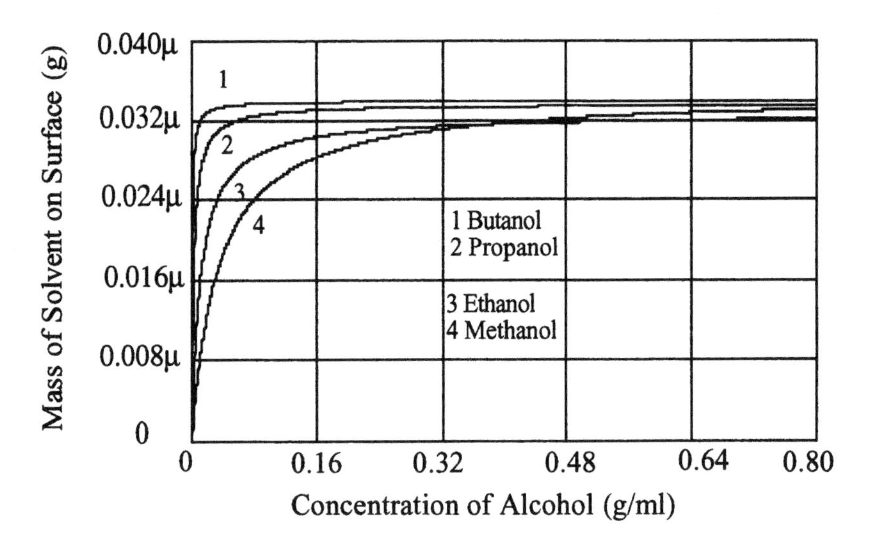

In contrast, the mono-layer of methanol is built up much more slowly and is not complete until the concentration of methanol in the aqueous mixture is about 35%w/v. The behavior of methanol on the reverse phase is reminiscent of the adsorption of chloroform on the strongly polar silica gel surface. The complementary nature of the silica gel surface and that of the reverse phase is clearly apparent. It is also clear that strongly dispersive solvents might form bi-layers on the reverse phase surface just as polar solutes form bi-layers on the highly polar surface of silica gel. In fact, to date there has been no experimental evidence furnished that would support the formation of bi-layers on the surface of reverse phases, although their formation is likely and such evidence may well be forthcoming in the future.

Solutes will interact with the reverse phase surface in much the same way as they do with the silica gel surface. There will be basically two forms of interaction, by sorption and by displacement. Sorption interaction has been experimentally confirmed by Scott and Kucera (10) by measuring the adsorption isotherm of acetophenone on the reverse phase RP18 from a 40%w/v acetonitrile mixture in water. The authors noted that there was *no* change in the acetonitrile concentration, as the solute was adsorbed. Displacement interactions, although certain to occur, do not appear to have been experimentally demonstrated to date.

Paired Ion Reagents

Although ion exchange bonded phases have been produced and used effectively in ion exchange chromatography, they are not very popular, as they tend to be unstable at a pH below 4.0 and above 8.0. Even within that pH range, in the presence of significant salt concentrations, they still have relatively poor long term stability. Furthermore, as ionic materials can not enter the pores of the bonded phase due to ion exclusion, it is sometimes difficult to organize a sufficiently large interactive surface to retain and resolve ionic solutes. In an attempt to address this problem, the so called *Paired Ion Reagents* were introduced. The separation technique using these reagents was given the title, somewhat vaingloriously, Paired Ion Chromatography (PIC).

Ion pair reagents may be anion or cation exchangers such as tertiary butyl ammonium iodide, or hexane sulfonate, that can be added to the mobile phase in small quantities and which can affect the retention of ionic materials in two ways. In the first instance, at low solvent concentrations the aliphatic groups of PIC reagents will interact strongly with the reverse phase and form a layer of ion exchange material on the surface. This mono-layer formation of aliphatic ion exchange material is similar to the layer of butanol depicted by the first curve in figure 9. It follows that, for a six carbon chain aliphatic sulfonate, the mono-layer will be completed at concentrations of less than 1%w/v of the reagent in the mobile phase.

It must be kept in mind, however, that the PIC reagents themselves will not be able to enter the pores of the reverse phase due to ion exclusion and will merely coat the external surface of the stationary phase. Nevertheless, under these circumstances ions will interact with the surface charges and materials will be retained and separated due to ionic interactions. However, at higher solvent concentrations, it has also been suggested that the PIC reagents work in an entirely different manner.

As the solvent concentration increases, the PIC reagents will interact more strongly with the mobile phase and will be less strongly adsorbed on the reverse phase surface. As a consequence, there will be less ion exchange material on the stationary phase surface. This is clearly demonstrated by the adsorption isotherm of octane sulfonate shown in figure 10.

Figure 10

The Adsorption Isotherm for Octane Sulfonate on a Reverse Phase

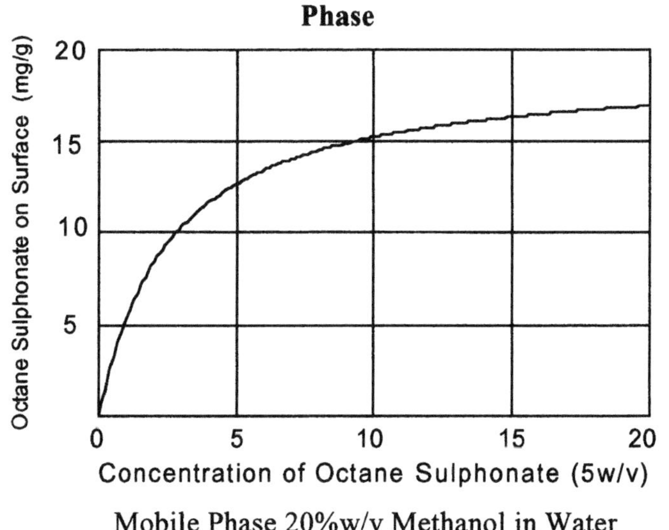

Mobile Phase 20%w/v Methanol in Water

It is seen that although the Langmuir form of the isotherm is maintained, the curve is much flatter and the sulfonate must be present at a concentration of betweent 10 to 15 %w/w for the mono-layer to be complete. Such levels of PIC reagents would be prohibitive in chromatography for many reasons.

However, although there is little ion pair reagent on the surface, at higher solvent concentration, the reagent can still assist in the separation of ionic materials. If the pH of the mobile phase is appropriately adjusted, the PIC reagent interacts with any counter ionic substances in the sample, forming a salt with a very weak dissociation constant. The ionic component of the sample, now an undissociated salt with a relatively long dispersive 'tail' can interact strongly with reverse phase and thus be retained.

As already stated, this can only occur if the pH of the mobile phase is adjusted so that the 'ion-pair' are *strongly* associated and thus can exist in the unionized form long enough to allow the normal chromatographic distribution processes to take place. It would appear that the precise manner in which the PIC reagents function remains a little uncertain, particularly at high solvent concentrations. More research is required before the retentive mechanism of the PIC reagents can be unambiguously identified. Nevertheless, PIC reagents can be used to advantage under certain circumstances to help separate difficult mixtures of ionic materials.

The Use of Bonded Phases in Elution Chromatography

Bonded phases are the most useful types of stationary phase in LC and have a very broad range of application. Of the bonded phases, the *reverse* phase is by far the most widely used and has been applied successfully to an extensive range of solute types. The reverse phases are commonly used with mobile phases consisting of acetonitrile and water, methanol and water, mixtures of both acetonitrile and methanol and water, and finally under very special circumstances tetrahydrofuran may also be added. Nevertheless, the majority of separations can be accomplished using simple binary mixtures.

Retention is reduced by increasing the solvent concentration, conversely, retention is increased by increasing the proportion of water. Binary mixtures of water and acetonitrile or water and methanol are, unfortunately, not simple binary mixtures because, as is well known, they associate strongly with one another. Thus, a nominally binary mixture of methanol and water is, in fact, a ternary mixture of water, methanol and water associated with methanol. It follows, that some discussion on aqueous solvent mixtures would be pertinent.

Aqueous Solvent Mixtures

The association of methanol and water was examined by Katz, Lochmüller and Scott (11) using volume change on mixing and refractive index data and established that the methanol/water solvent system was indeed a complex ternary system.

Figure 11

Diagram of the Ternary Solvent System for Methanol/Water Mixtures

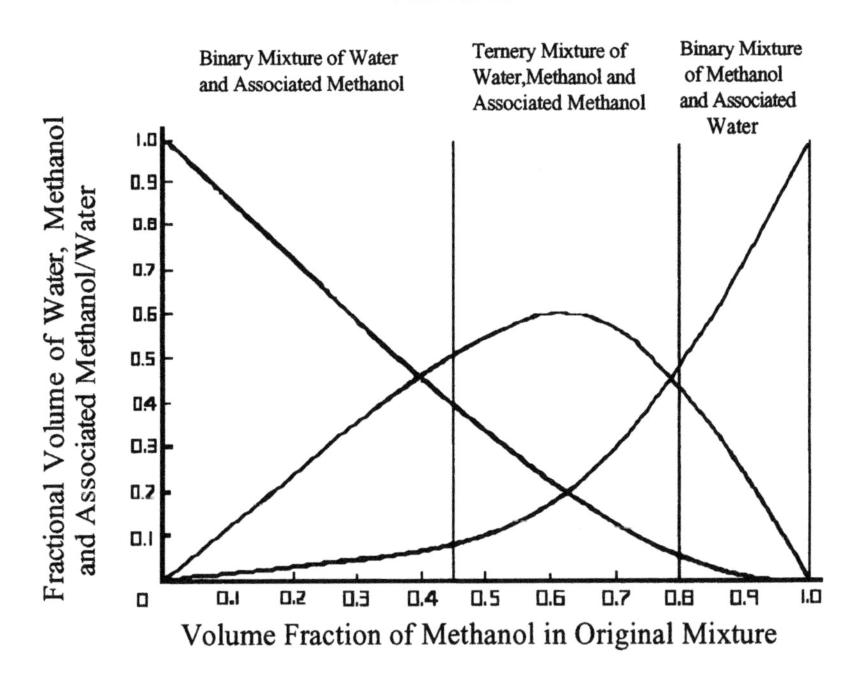

They identified the association equilibrium constant and from that, calculated the distribution of the different components of a methanol water mixture form zero to 100% methanol. The curves they obtained are shown in figure 11. It is seen that there are three distinct ranges of methanol concentration over which the effect of the solvent mixture on retention will be very different. From zero to 40%v/v of methanol in the original mixture, the solvent will behave as though it were a binary mixture of *water* and *methanol associated with water*. From 40%v/v to 80%v/v of methanol in the original mixture, the solvent will behave as though it were a ternary mixture of *water*, *methanol* and *water associated with methanol*. Finally from 80%v/v to 100%v/v of methanol in the original mixture the solvent will again behave as though it were again a binary mixture, but this time as a mixture of *methanol* and *water associated with methanol.*

It is the *solvent,* in this case methanol, that is responsible for reducing the distribution coefficients of the solutes with respect to the stationary phase (and consequently, their retention) by increasing the solute-solvent interactions in the mobile phase. Bearing this in mind, then the curves shown in figure 11 can explain some of the unique characteristics of methanol water mixtures when used as the mobile phase in reversed phase LC.

From figure 11 it is seen that when the original methanol mixture contains 50%v/v of methanol there is little free methanol available in the mobile phase to elute the solutes as it is mostly associated with water. Subsequently, however, the proportion of methanol *unassociated* with water increases rapidly in the solvent mixture which quickly reduces solute retention. In practice, when employing gradient elution, this rapid increase in methanol concentration over the latter part of the program, is accommodated by the use of a convex gradient profile. The *convex* gradient will compensate for the strongly *concave* form of the unassociated methanol concentration profile shown in figure 11. The strong association of methanol with water can also account for the fact that proteins can tolerate a significant amount of methanol in the mobile phase without them becoming denatured. This is because there is virtually no methanol present in the mixture which could cause protein denaturation, since all the methanol is in a deactivated state associated with water.

Katz, Lochmüller and Scott also examined acetonitrile/water and tetrahydrofuran (THF)/water mixtures using a similar procedure and showed that there was significant association between the water and both solvents but not to the same extent as methanol/water. At the point of maximum association for methanol, the solvent mixture contained nearly 60% of the methanol/water associate. In contrast the maximum amount of THF associate that was formed amounted to only about 17% and for acetonitrile the maximum amount of associate that was formed was as little as 8%. It follows that acetonitrile water mixtures would be expected to behave more nearly as simple binary mixtures than methanol/water or THF/water mixtures.

The Use of Macro-Porous Polymers as Stationary Phases

The effective use of polymeric ion exchange materials for chromatographic purposes started to become established with the introduction of macro-porous polymers in the early sixties (12-14). The essential technical advance lay in the macro-porous nature of the resin packing, which consisted of resin particles a few microns in diameter, which in turn were comprised of a fused mass of polymer micro-spheres a few Angstroms in diameter. The structure is very similar to the silica gel particles, which consist of a fused mass of *primary* particles of polymeric silicic acid. The resin polymer micro-spheres play the same part as the silica gel primary particles, and confer on the polymer a relatively high surface area as well as high porosity. The high surface area was the key to the improvement, as it provided increased solute retention and selectivity together with a superior loading capacity which in turn provided a larger quantitative dynamic range of analysis. Chemically, the material consisted of a highly cross-linked polystyrene resin with about a 50Å pore diameter and for ion exchange chromatography, ionogenic groups of appropriate charge were chemically attached (for example, by sulfonation). The early materials manufactured by Rohm and Haas were the Amberlite copolymers that had a pore volume between 0.4 and 0.5 ml, particle sizes of 300 to 1200 microns, a mean pore diameter of about 235 Å and a surface area of about 160 sq m per g. The first materials manufactured by Rom and Haas were for production purposes and thus, had the large particle diameters which were unsuitable for analytical separations. Today there are a number of manufactures that produce resin type stationary phases with particle diameters as small as 2.5 or 3.0 microns.

The more popular resin based packings are based on the co-polymerization of polystyrene and divinylbenzene. The extent of cross-linking determines its rigidity and the greater the cross-linking the harder the resin becomes until, at extremely high cross-linking, the resin becomes brittle. In order to produce an ion exchange resin the surface of the polystyrene-divinyl benzene copolymer is finally reacted with appropriate reagents and covered with the required ionogenic interacting groups.

The pore structure of most cross-linked polystyrene resins are the so called macro-reticular type which can be produced with almost any desired pore size, ranging from 20Å to 5,000Å. They exhibit strong dispersive type interaction with solvents and solutes with some polarizability arising from the aromatic nuclei in the polymer. Consequently the untreated resin is finding use as an alternative to the C8 and C18 reverse phase columns based on silica. Their use for the separation of peptide and proteins at both high and low pH is well established.

An example of the efficacy of the resin phases used as an alternative to a conventional silica based reverse phase is shown in figure 12 where the separation of the three tocopherols are shown separated on the C18 Polymer Column and The ODA-A 120Å silica gel based columns. The columns were 15 cm long, 4.6 mm i.d., operated at a flow rate of 0.5 ml/min at 30°C with a mobile phase of 98% methanol/2% water.

Figure 12

Comparison of the Separation of Some Tocopherols on a Polymer C18 Column and a Conventional Silica Based ODS Column

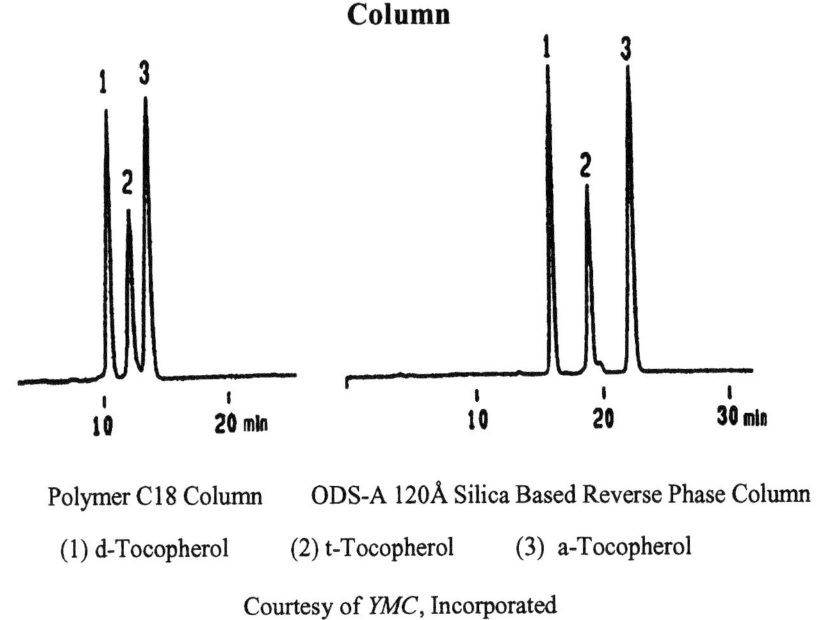

Polymer C18 Column ODS-A 120Å Silica Based Reverse Phase Column

(1) d-Tocopherol (2) t-Tocopherol (3) a-Tocopherol

Courtesy of *YMC*, Incorporated

It is seen that the polymer resin does not have the same retentive capacity as the conventional reverse phase column and thus, will not exhibit the same resolution or the equivalent loading capacity. Nevertheless, the polymer column will function over a wide range of pH whereas the silica based columns will be restricted to operating within a pH of 4.0 to 8.0 at the most.

Despite this relatively new use of polymer stationary phases, ion exchange chromatography is still their major application area. One of the problems involved with ion exchange chromatography is the presence of the buffer or salt ions as a background conductivity signal to the conductivity detector. The performance of conductivity detectors will be discussed with detectors generally in a later chapter and in the mean time, it will be sufficient to say that the ions producing the background conductivity can be selectively removed by a separate column or membrane system situated between the analytical column and the detector. This technique is called ion suppression and, although effective, further band dispersion occurs in the suppression system which can significantly reduce the column efficiency and impair the resolution. An example of the separation of alkali metal cations employing an ion exchange resin with a membrane suppression system is shown in figure 13.

Figure 13

Separation of Some Alkali Metals Using an Ion Exchange Resin in Conjunction with a Cation Micro-Membrane Ion Suppressor

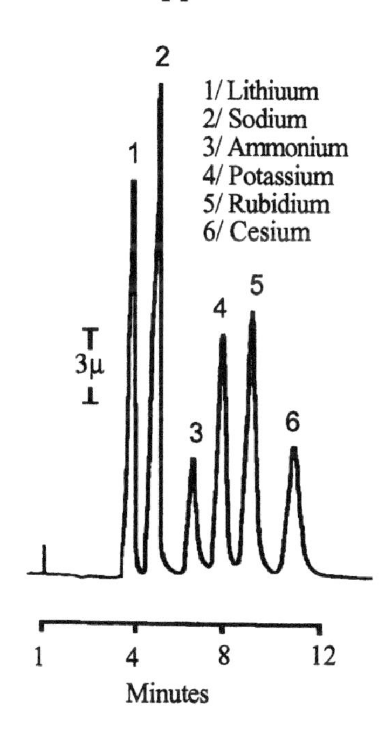

Courtesy of DIONEX Inc.

The column was packed with IonPac CS3 ion exchange resin and operated at a flow-rate of 1.0 ml per minute. The mobile phase gradient ranged from 0.04 M HCl to 0.055 M HCl. The composition of the sample was Li^+ (2.3 ppm), Na^+ (10 ppm), $NH4^+$ (10 ppm), K^+ (10 ppm), Rb^+ (30 ppm),and Cs^+ (30 ppm).

Although porous resins can provide higher loading capacity and thus, greater solute retention and selectivity than their non-porous counterparts, non-porous ion exchange resins can be very useful in analytical separations due to their potential for producing higher column efficiencies. Furthermore, due to the natural phenomena of 'ion exclusion' from stationary phase pores, it is unlikely that any ion exchange moieties inside the pores can take part in solute retention anyway. TOSOHAAS has produced a number of non-porous ion exchange resins that have particle sizes of 2.5μ in diameter. However, such columns appear to give little more than about 20,000 theoretical plates per meter. If a 'plate height' of 5 microns was obtained (the minimum plate height that can be obtained theoretically from a well packed column is twice the particle diameter), then a well packed column should provide about 200,000 theoretical plates per meter. It follows that the efficiencies that are obtainable from silica based materials have not yet been achieved.

Figure 14

The Separation of a Protein Mixture on a Non-Porous Ion Exchange Resin

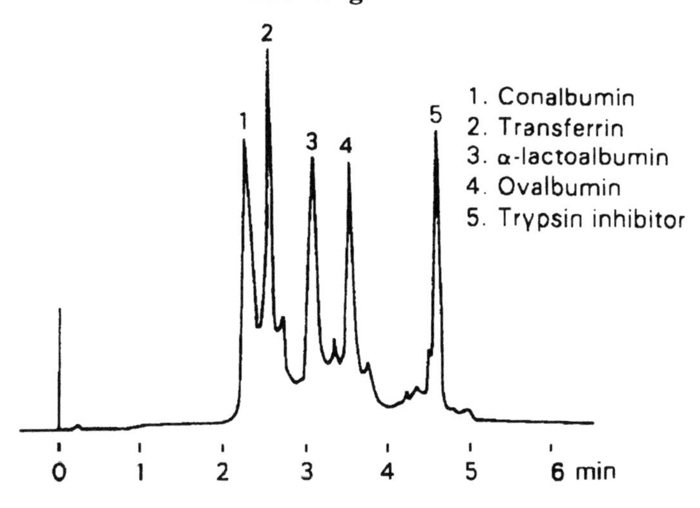

Courtesy of TOSOHAAS

Nevertheless, as indicated by the separation shown in figure 14, the efficiencies that have been obtained can be extremely useful. Figure 14 shows the separation of a protein mixture in 6 minutes. The column was 35 cm long, 4.6 mm i.d. and packed with a non-porous resin designated as TSKgel DEAE-NPR and operated at 1.5 ml / min with a gradient from 20 mM tris-HCl buffer (pH 8.0) to the same solvent containing 0.5 M sodium chloride in 10 minutes. It is seen that a fast, relatively clean separation is obtained at a pH at which silica based bonded phases would certainly show some instability. As already stated, polystyrene cross-linked with divinyl benzene in macro-reticular form, can be made with a wide rang of pore sizes. The material being largely dispersive in character, if used with a sufficiently dispersive solvent, the dispersive interactions in both phases can be made equal and thus separate solutes solely on the basis of size exclusion. This type of application of the resin media is becoming increasingly popular particularly in the biotechnology field. It has been found particularly valuable in the separation of soluble organic polymers.

Figure 15

The Separation of a Peptide Mixture on TSK Gel Octadecyl-NPR

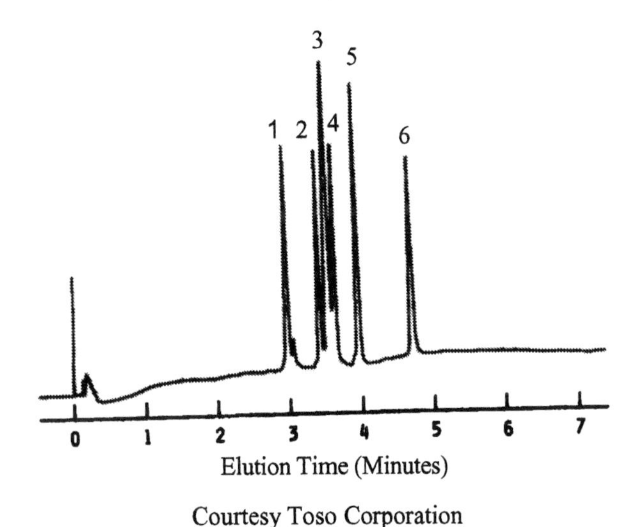

Courtesy Toso Corporation

The basic polymer appears to be a hydroxylated polyether to which octadecyl chains have been bonded and so it behaves as a reverse phase exhibiting dispersive interactions with the solutes. An example of the separation of a series of peptides is shown in figure 15. The column was 3.5 cm long, 4.6 mm i.d. The solutes shown were (1) α-endorphin, (2) bombesin, (3) γ-endorphin, (4) angiotensin, (5) somatostatin and (6) calcitonon. The separation was carried out with a 10 min linear program from water containing 0.2% trifluoroacetic acid to 80% acetonitrile.

Polymer Stationary Phases in Elution Chromatography

In addition to the cross-linked polystyrene material, macro-reticular or porous resin beads have been prepared from polymethylmethacrylate beads which are claimed to be hydrophilic (polar and will interact with water) in nature. Polyacrylamide has also been processed into bead form for use as LC stationary phases. There appears to be a new type of organic polymeric stationary phase produced almost every week and newcomers to the field might well become bemused by the wide choice available. However, if it is understood that the basis of retention by all stationary phases depends on ionic, polar or dispersive interactions between it and the solute then the novelty of the product and its value can be rationally examined. However recent and exotic a new stationary phase may appear to be, its chromatographic performance can probably be exactly simulated by a silica based reverse phase with an appropriately chosen solvent system. For this reason the precise chemical nature of the stationary phase needs to be known. Such information should be forthcoming from all reputable manufacturers. The advantages of the polymer type phases will be where the mixture needs to be separated at pH values outside the stability range of silica based phases and where the activity of the phase can cause deconformation of biologically active materials. These advantages will apply to the use of polymer stationary phase both for exclusion and elution chromatography. It is rare that the chemical nature of a stationary phase makes it uniquely suitable for a particular analysis.

General Advice for the Unknown Sample

The newcomer to chromatography, faced with a hitherto unknown sample, would do well to start with a C8 silica based reverse phase and an acetonitrile water mixture as a mobile phase and carry out a gradient elution from 100% water to 100% acetonitrile. From the results, the nature and the complexity of the sample can be evaluated and a more optimum phase system can be inferred.

References

1/ "*Silica Gel and Bonded Phases*", R. P.W Scott, John Wiley and Sons, Ltd. Chichester, England (1993).

2/ K. K. Unger, *Angew . Chem . Int . Engl .*, **11**(1972)267.

3/ M. Le Page, R. Beau, and J. Duchene, *Fr. Pat.*, **No. 1.473,240** (1967).

4/ M. Le Page and A. de Vries, *Fr. Pat.*, **No. 1.475,929** (1967).

5/ R.P.W. Scott and P. Kucera, *J. Chromatogr.* **149**(1978)93.

6/ I. Halasz and I. Sebastian, *Angew Chem. Interact. Ed.*, **8**(1969)453.

7/ J. J. Kirkland and J. J. DeStefano, *J. Chromatogr. Sci.,* **8**(1970)309.

8/ J.H Knox, K.K. Unger and H. Mueller, *J.Liq.Chromatogr.*, **6**(1983)1.

9/ R. P. W. Scott and C. F. Simpson, *Farady Discussions of the Royal Society of Chemistry*, **Faraday Symposium, No. 25** (1980) 69.

10/ R. P. W. Scott and P. Kucera, *J. Chromatogr.*,**142**(1977)213.

11/ E. D. Katz, C .H. Lochmüller and R. P . W.Scott, *Anal. Chem.,* **61**(1981)344.

12/ R. Kunin, E. F. Meitzner and N. Bostnick, *J. Am. Chem.Soc.,* **84**(1962)84.

13/ J. R. Millar, D. G. Smith, E. E. Marr and T. R. E. Kressman, *J. Chem. Soc.,* **218**(1963).

14/ R. Kunin, E. F. Meitzner, J. A. Oline, S. A. Fischer and N.W. Frisch, *Ind. Eng. Chem. Prod. Res. Develop.,* **1**(1962)140.

15/ YOSHIO KATO, SHIGERU NAKATANI,TAKASHI KITAMURA, YOSUKE YAMASAKI and TSUT HASHIMOTO, *J. Chromatogr.*, **502**(1990)416.

4
The Liquid Chromatography Column

It has been shown that, in LC, the size of the distribution coefficient of a solute between the two phases determines the extent of its *retention*. As a consequence, the *difference* between the distribution coefficients of two solutes establishes the extent of their *separation*. The distribution coefficients are controlled by the nature and strength of the molecular interactions that takes place between the solutes and the two phases. Thus it is the choice of the *phase system* that primarily determines the separation that is achieved by the chromatographic system.

At this point, it is important to stress the difference between *separation* and *resolution*. Although a pair of solutes may be separated they will only be *resolved* if the peaks are kept sufficiently narrow so that, having been moved apart (that is, separated), they are eluted discretely. Practically, this means that firstly there must be sufficient stationary phase in the column to move the peaks apart, and secondly, the column must be constructed so that the individual bands do not spread (disperse) to a greater extent than the phase system has separated them. It follows that the factors that determine peak dispersion must be identified and this requires an introduction to the Rate Theory. The Rate Theory will not be considered in detail as this subject has been treated extensively elsewhere (1), but the basic processes of band dispersion will be examined in order to understand the function of the column, column design and how to choose an appropriate column for a particular application.

The Rate Theory

The Rate Theory provides an equation that allows the variance of an eluted peak (the square of half the peak width) to be calculated from the various physical properties of the two phases, the solute and the packing. This is called the HETP equation and the reasons for this will become clear later. A number of HETP equations have been developed (2-5) but the one that has received the greatest experimental support is that developed by Van Deemter *et al* in 1956 (6). Van Deemter *et al* proposed that the overall spreading of a peak was the result of a number of individual, non-interacting, random processes that, taken together, accounted for the total peak dispersion. The dispersion processes that takes place in a chromatographic column are, indeed, random and non-interacting, and do meet this requirement. It is not possible to sum the band *widths* resulting from each individual dispersion process but, due to the random nature of the dispersion processes, it is possible to sum all the respective *variances*. This procedure is called the Summation of Variances.

The Summation of Variances

Assuming there are (p) non-interacting, random dispersive processes occurring in the chromatographic system, then any process (q) acting alone will produce a Gaussian curve having a variance σ_q^2.

$$\text{Hence,}\quad \sigma_1^2+\sigma_2^2+\sigma_3^2+\ldots+\sigma_q^2+\ldots+\sigma_p^2 = \sigma^2 \qquad (1)$$

where σ^2 is the total variance of the solute band.

Equation (1) is the algebraic expression of the principle of the summation of variances. If the individual dispersion processes that take place in a column can be identified, and the variance that results from each dispersion determined, then the variance of the final band can be calculated from the sum of all the individual variances. This is how the Rate Theory provides an equation for the final variance of the peak leaving the column.

Before progressing to the Rate Theory Equation, an interesting and practical example of the use of the summation of variances is the determination of the maximum sample volume that can be placed on a column. This is important because excessive sample volume broadens the peak and reduces the resolution. It is therefore important to be able to choose a sample volume that is as large as possible to provide maximum sensitivity but, at the same time insufficient, to affect the overall resolution.

The Maximum Sample Volume

As already stated any sample placed on a column will have a finite volume, and the variance of the injected sample will contribute directly to the final peak variance. It follows that the maximum volume of sample that can be placed on the column must be limited, or the column efficiency will be seriously reduced. Consider a volume V_i, injected onto a column. Normal LC injections will start initially as a rectangular distribution and the variance of the eluted peak will be the sum of the variances of the injected sample plus the normal variance of the eluted peak.

Thus,

$$\sigma^2 = \sigma_i^2 + \sigma_c^2$$

where σ^2 is the variance of the eluted peak,
σ_i^2 is the variance of the injected sample,
and σ_c^2 is the variance due to column dispersion.

It is generally accepted that the maximum increase in band width that can be tolerated due to any extraneous dispersion process is a 5% increase in standard deviation (or a 10% increase in peak variance).

Let a volume (V_i) be injected onto a column resulting in a rectangular distribution of sample at the front of the column. According to the principle of the Summation of Variances, the variance of the final peak will be the sum of the variances of the sample volume plus the normal variance of a peak for a small sample.

Now the variance of the rectangular distribution of a sample, volume (V_i), is $\left(\frac{V_i^2}{12}\right)$ and, from the Plate Theory, $\sigma_v^2 = \left(\sqrt{n}(v_m + Kv_s)\right)^2$.

Then, assuming a 5% increase in peak width due to the sample volume,

$$\frac{V_i^2}{12} + \left(\sqrt{n}\right)(v_m + Kv_s)^2 = \left(1.05\sqrt{n}\,(v_m + Kv_s)\right)^2$$

Thus,

$$\frac{V_i^2}{12} = n\,(v_m + Kv_s)^2\left(1.05^2 - 1\right)$$

$$= n(v_m + Kv_s)^2\,0.102$$

Consequently,

$$V_i^2 = n\,(v_m + Kv_s)^2 1.23$$

or

$$V_i = \sqrt{n}\,(v_m + Kv_s)\,1.1$$

Now from the Plate Theory given in chapter 2, $V_r = n\,(v_m + Kv_s)$

Then,

$$V_i = \frac{1.1\,V_r}{\sqrt{n}} \qquad (2)$$

It is seen that the that the maximum sample volume that can be tolerated can be calculated from the retention volume of the solute concerned and the efficiency of the column. However, a sample does not consist of a single component, and it is therefore important that the resolution of any solute, irrespective of where it is eluted, is not excessively dispersed by the sample volume. The solute that will be most affected will be the solute eluted in the smallest volume, which will be the solute eluted close to the dead volume. Consequently for general use the dead volume should be used in equation (2) as opposed to the retention volume of a specific solute.

Thus equation (2) becomes $V_i = \frac{1.1 V_0}{\sqrt{n}}$

Now the total volume (V_C) of a column radius (r) and length (l) will be $\pi r^2 l$. Furthermore, the volume occupied by the mobile phase will be approximately $0.6V_C$ (60% of the total column volume is occupied by mobile phase). Thus as a general rule the maximum sample volume that can be employed without degrading the resolution of the column is

$$V_i = \frac{0.60\,\pi r^2 l}{\sqrt{n}} \qquad (3)$$

From a practical point of view it is better to use the maximum volume of sample possible, assuming there is no mass overload. This will allow the detector to be operated at the lowest possible sensitivity and in doing so provide the greatest detector stability and, as a consequence, the highest accuracy.

The analyst can calculate the maximum sample volume (V_i) in a simple manner from the efficiency of a peak eluted close to the dead volume and the dimensions of the column.

The Van Deemter Equation

Van Deemter considered peak dispersion results from four spreading processes that take place in a column, namely, the Multi-Path Effect, Longitudinal Diffusion, Resistance to Mass Transfer in the Mobile Phase and Resistance to Mass Transfer in the Stationary Phase. Each one of these dispersion processes will now be considered separately and an expression for their contribution to the variance per unit length of column established. The variance per unit length of the column for all the dispersion processes will be obtained from a sum of the individual variance contributions.

The Multi-Path Effect

In a packed column the solute molecules will describe a tortuous path through the interstices between the particles and it is obvious that some will travel shorter paths than the average and some longer paths. The multi-path effect is illustrated in figure 1.

Figure 1

Dispersion by the Multi-Path Effect

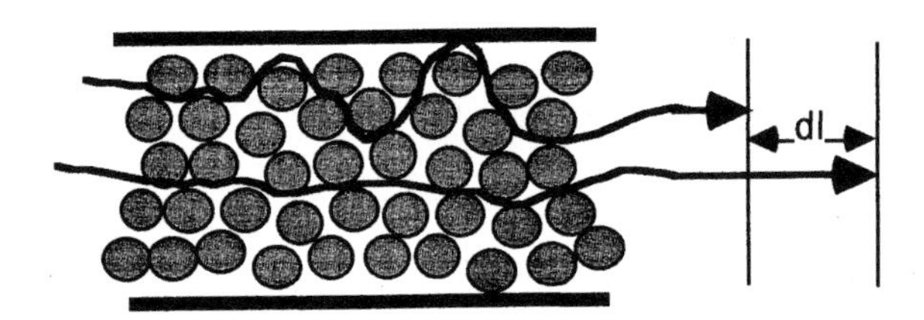

Consequently, some molecules, those taking the shorter paths, will move ahead of the average and some, those that take the longer paths, will lag behind (dl) thus causing band dispersion.

Van Deemter *et al* derived the following function for the variance per unit length contribution of the multi-path effect (σ_M^2) to the overall variance of the column,

$$\sigma_M^2 = 2\,\lambda\, d_p \qquad (4)$$

where (d_p) was the particle diameter of the packing,
and (λ) was a constant that depended on the quality of the packing.

Longitudinal Diffusion

If a local concentration of solute is placed at the midpoint of a tube filled with either a liquid or a gas, the solute will slowly diffuse to either end of the tube. It will first produce a Gaussian distribution with a maximum concentration at the center and finally, when the solute reaches the end of the tube, 'end' effects occur and the solute will continue to diffuse until there is a constant concentration throughout the length of the tube. This diffusion effect occurs in the mobile phase of a packed LC column but the end effects are never realized. The diffusion process is depicted in figure 2.

Figure 2

Peak Dispersion by Longitudinal Diffusion

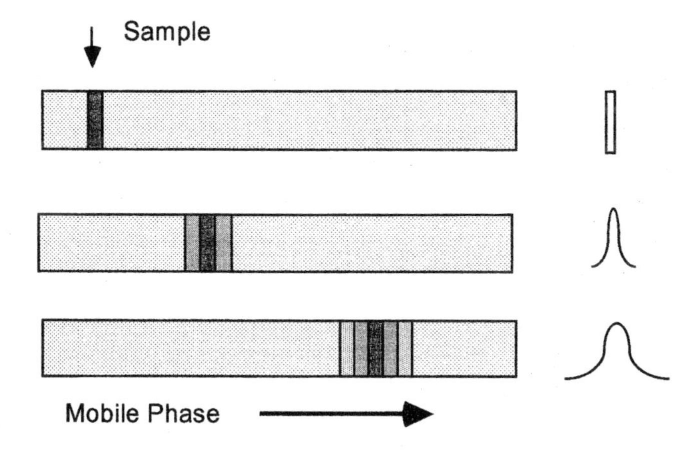

The dispersion described in figure 2 shows that the longer the solute band remains in the column, the greater will be the extent of longitudinal diffusion. Since the length of time the solute remains in the column is inversely proportional to the mobile phase velocity, so will the dispersion be inversely proportional to the mobile phase velocity. Van Deemter *et al* derived the following expression for the variance/unit length contribution due to longitudinal diffusion, (σ_L^2), to the overall variance/unit length of the column.

$$\sigma_L^2 = \frac{2\,\gamma\,D_m}{u} \qquad (5)$$

where (D_m) is the diffusivity of the solute in the mobile phase,
(u) is the linear velocity of the mobile phase,
and (γ) was a constant that depended on the quality of the packing.

The Resistance to Mass Transfer in the Mobile Phase

During the movement of a solute band along a column, the solute molecules are continually transferring from the mobile phase into the stationary phase and back from the stationary phase into the mobile phase. This transfer process is not instantaneous, because a finite time is required for the molecules to traverse (by diffusion) through the mobile phase in order to reach the interface and enter the stationary phase. Thus, those molecules close to the stationary phase will enter it almost immediately, whereas those molecules some distance away from the stationary phase will find their way to it a significant interval of time later. However, since the mobile phase is moving, it is obvious that during this time interval, those molecules that remain in the mobile phase will be swept along the column and dispersed away from those molecules that were close and entered the stationary phase immediately. The dispersion resulting from the resistance to mass transfer in the mobile phase is depicted in figure 3. The diagram depicts 6 solute molecules in the mobile phase, and those closest to the surface, 1 and 2, immediately enter the stationary phase. During the period while molecules 3 and 4 diffuse through the mobile phase to the interface, the mobile phase moves on. Thus, when molecules 3 and 4 reach the interface they will enter the stationary phase some distance ahead of the first two. Finally, while molecules 5 and 6 diffuse to the interface the mobile phase moves even further down the column until molecules 5 and 6 enter the stationary phase further ahead of molecules 3 and 4. Thus, the 6 molecules, originally relatively close together, are now spread out in the stationary phase. This explanation is a little over-simplified but gives a correct description of the mechanism of mass transfer dispersion.

Figure 3

Resistance to Mass Transfer in the Mobile Phase

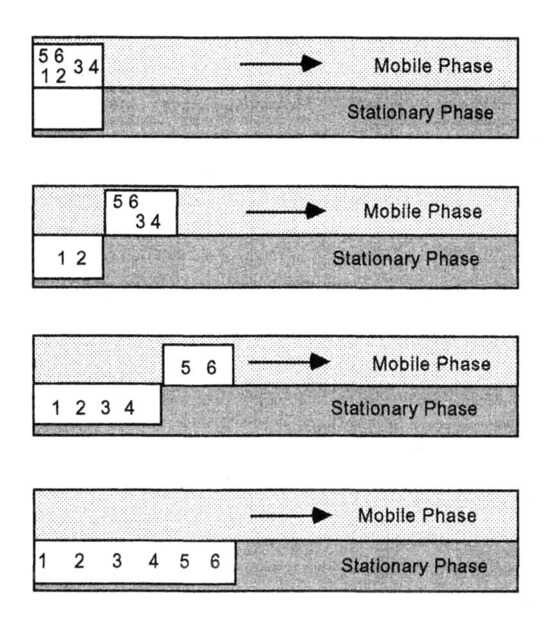

Van Deemter derived an expression for the contribution to variance/unit length by the resistance to mass transfer in the mobile phase, ($\sigma^2_{R\,M}$), which is given as follows:

$$\sigma^2_{R\,M} = \frac{f_1(k')d_p^2}{D_m}u \qquad (6)$$

where (k') is the capacity ratio of the solute, and the other symbols have the meaning previously ascribed to them.

The Resistance to Mass Transfer in the Stationary Phase

The dispersion resulting from the resistance to mass transfer in the stationary phase is exactly analogous to that in the mobile phase. Those solute molecules close to the surface of the stationary phase will leave the surface and enter the mobile phase a significant time before those that have diffused farther into the stationary phase and have a longer distance to diffuse back to the surface. Thus, as those molecules that were close to the surface will be swept along by the moving phase, they will be dispersed from those molecules still diffusing to the surface. The dispersion resulting from the resistance to mass transfer in the stationary phase is depicted in figure 4.

Figure 4

Resistance to Mass Transfer in the Stationary Phase

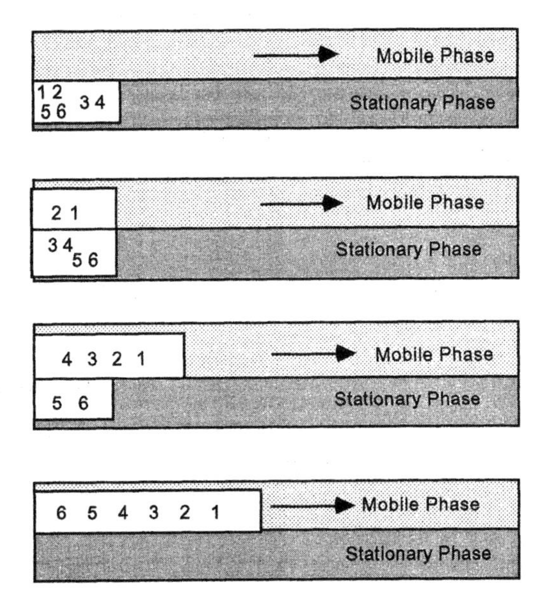

At the start, again with six molecules in the stationary phase, molecules 1 and 2, the closest to the surface will enter the mobile phase and begin moving along the column. This movement will continue while molecules 3 and 4 diffuse to the interface at which time they will enter the mobile phase and start following molecules 1 and 2. All four molecules will continue their journey along the column while molecules 5 and 6 diffuse to the mobile phase/stationary phase interface. By the time molecules 5 and 6 enter the mobile phase the other four molecules will have been smeared along the column and the original 6 molecules will have suffered dispersion. Van Deemter derived an expression for the contribution to variance/unit length by the resistance to mass transfer in the stationary phase,($\sigma^2_{R\,S}$), which is as follows:

$$\sigma^2_{R\,S} = \frac{f_2(\,k'\,)\,d^{\,2}_{f}}{D_S}\,u \qquad (7)$$

where (k') is the capacity ratio of the solute,
(df) is the effective film thickness of the stationary phase,
(D_S) is the diffusivity of the solute in the stationary phase,
and the other symbols have the meaning previously ascribed to them.

Now, applying the law of Summation of Variances,

$$\sigma^2_x = \sigma^2_M + \sigma^2_L + \sigma^2_{R\,M} + \sigma^2_{R\,S} \qquad (8)$$

where (σ^2_x) is the total variance/unit length of the peak.

Thus substituting for (σ^2_M), (σ^2_L), ($\sigma^2_{R\,M}$) and ($\sigma^2_{R\,S}$) from equations (4), (5), (6) and (7), respectively,

$$\sigma^2_x = 2\,\lambda d_p + \frac{2\,\gamma\,D_m}{u} + \frac{f_1(k')d^2_p}{D_m}\,u + \frac{f_2(\,k'\,)\,d^{\,2}_{f}}{D_S}\,u \qquad (9)$$

Equation (9) is the basic form of the Van Deemter equation that describes the variance per unit length of a column in terms of the physical properties of the column contents and the distribution system.

Now if (σ_l) is the peak width of a solute having passed though a column of length (L) which is a *retention distance* (L) and the *total variance* of the peak will be (σ^2_L).

Then, by simple proportion

$$\frac{\text{Length Variance}}{\text{Retention Distance}^2} = \frac{\text{Volume Variance}}{\text{Retention Volume}^2}$$

or

$$\frac{\sigma_L^2}{L^2} = \frac{\sigma_v^2}{V_r^2}$$

Now, from the Plate Theory that was discussed in the previous chapter,

$$V_r = n(v_m + Kv_s) \text{ and } \sigma_v = \sqrt{n}(v_m + Kv_s)$$

Thus,

$$\frac{\sigma_L^2}{L^2} = \frac{(\sqrt{n}(v_m + Kv_s))^2}{(n(v_m + Kv_s))^2} = \frac{1}{n}$$

$$\text{or} \quad \frac{\sigma_L^2}{L} = \frac{L}{n}$$

Now,

$$\frac{\sigma_L^2}{L} = \sigma_x^2 \text{, the variance per unit length of the column,}$$

and $\frac{L}{n}$ is the height (length) equivalent to a theoretical plate or HETP.

Consequently, (σ_x^2) the variance per unit length of a column is numerically equal to the height equivalent to a theoretical plate (H) and thus, equation (9) becomes

$$H = 2\lambda d_p + \frac{2\gamma D_m}{u} + \frac{f_1(k')d_p^2}{D_m}u + \frac{f_2(k')d_f^2}{D_S}u \qquad (10)$$

Hence the term "HETP equation" for the equation for the variance per unit length of a column.

The Significance of the HETP Equation

Equation (10) can be put into a simpler form:

$$H = A + \frac{B}{u} + (C_m + C_s)u \qquad (11)$$

where

$$A = 2\lambda d_p \quad B = 2\gamma D_m \quad C_m = \frac{f_1(k')d_p^2}{D_m} \text{ and } C_s = \frac{f_2(k')d_f^2}{D_S}$$

Equation (11) is that which describes the 'HETP Curve', or the curve that relates the variance per unit length or HETP of a column to the mobile phase linear velocity. A typical curve is shown in figure 5.

Figure 5

A Typical HETP Curve

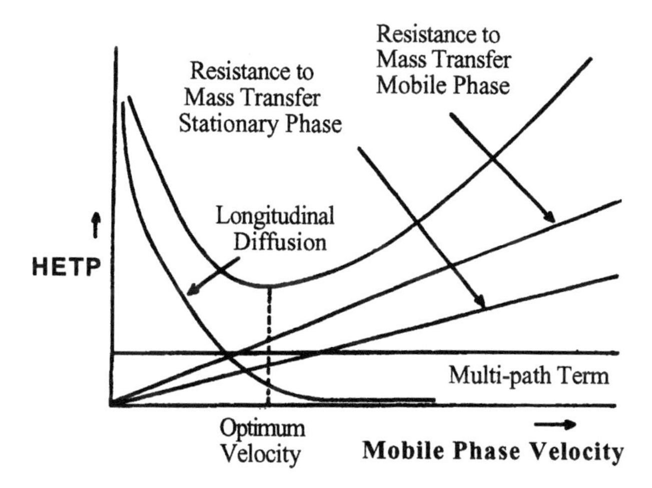

It is seen that the composite curve is hyperbolic, consisting of the sum of a constant term value (A) that describes the multi-path effect, a reciprocal function $\frac{B}{u}$ that describes the longitudinal diffusion and a linear function $(C_m + C_s)u$, which accounts for the dispersion arising from the two resistance to mass transfer effects.

The curve exhibits a minimum, which means that there is an optimum mobile phase velocity at which the column will give the minimum HETP and consequently a maximum efficiency. In practice this usually means that reducing the flow rate of a column will increase the efficiency and thus the resolution. In doing so, however, the analysis time will also be increased. As seen in figure 5, however, there is a limit to this procedure, as reducing the column flow rate so that the mobile phase velocity falls below the optimum will result in an *increase* in the HETP and thus a *decrease* in column efficiency.

It is seen that equation (10) is explicit and would permit the design of a column to have the particular efficiency required for a specified analysis. Column design, however, is an involved procedure and will not be described here; those interested will find column design discussed in detail in reference (1). Nevertheless equation (10) clearly indicates how high efficiencies can be obtained from LC columns. Not surprisingly, the major factor affecting the HETP or column efficiency (as predicted by Martin in 1941) is the particle size. The smaller the particle, the smaller will be the multi-path term and the smaller the resistance to mass transfer in mobile phase and thus, the smaller the overall HETP and the higher the column efficiency. The HETP equation and the HETP curve can also assist in the general operation of a chromatograph but firstly, the concept of the Reduced Chromatogram must be introduced.

The Reduced Chromatogram

The chromatographic challenge associated with a particular sample can be summarized by the *reduced chromatogram.* The reduced chromatogram consists of three peaks and is represented in figure 6.

Figure 6

The Reduced Chromatogram

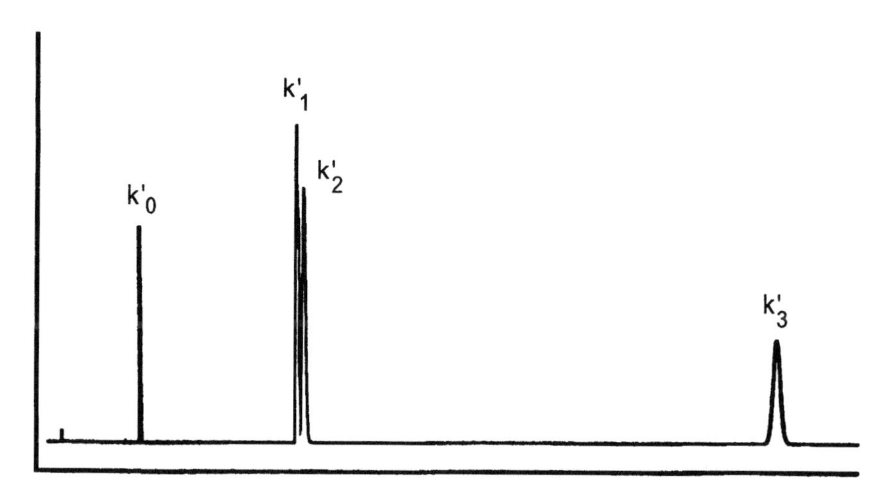

All samples can be reduced to such a chromatogram and if the reduced chromatogram can be resolved then, almost without exception, so can the sample. In the following discussion it is assumed that all the components of the mixture have equal importance and must be isolated and quantitatively estimated. The analyst will, at times, be presented with samples for which a full analysis is not required and such samples will be discussed subsequently.

The time that the last peak is eluted in the reduced chromatogram represents the total analysis time after which, the analysis can be terminated. The two peaks that are eluted closest together are the most difficult to separate and which, for obvious reasons, are called the "critical pair".

The columns must be designed or chosen such that the critical pair are separated and, as a second priority, the last peak must be eluted in a reasonable time. The first peak in the chromatogram is not considered part of the reduced chromatogram and is included as the dead volume marker from which the capacity factors of each solute can be calculated, together with the separation ratio of the critical pair.

The above considerations apply to samples where all the components are of interest and all need to be separated and quantitatively assessed. In practice, for many samples, only specific components of the mixture are important and only those need to be separated from the matrix and be analyzed. The components of the matrix need not be resolved and they are of no interest. It follows, that under these circumstances the critical pair will be comprised of the component of interest that has the closest neighbor and the neighbor itself. Such a situation usually greatly simplifies the separation problem but it should be noted that the last peak must still be eluted before the next analysis can be carried out and so the analysis time may not be significantly reduced.

Resolution

The column that is required for an analysis must have the efficiency necessary to resolve the critical pair. However, it is necessary to decide what constitutes resolution, before the column efficiency can be calculated. How narrow must the peaks be maintained relative to their separation to permit an accurate quantitative analysis? In figure 7, five pairs of peaks are shown, the area of the smaller peak being half that of the larger peak.

Figure 7

Peaks Showing Different Degrees of Resolution

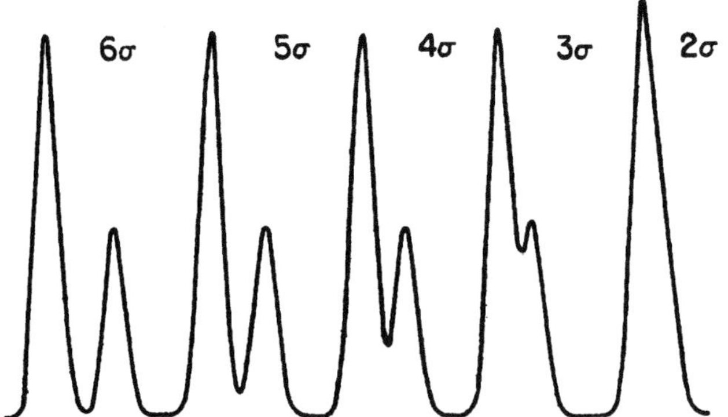

The mixture is identical in each example. The peaks are shown separated by 2, 3, 4, 5 and 6 (σ) and it is clear that a separation of 6σ would appear to be ideal for accurate quantitative results. Such a resolution, however, will often require very high efficiencies which will be accompanied by very long analysis times. Furthermore, a separation of 6σ is not *necessary* for accurate quantitative analysis. Even with manual measurements made directly on the chromatogram from a strip chart recorder, accurate quantitative results can be obtained with a separation of only 4σ. That is to say that duplicate measurements of peak area or peak height should not differ by more than 2%. (A separation of 4σ means that the distance between the maxima of the two peaks is equal to twice the peak widths). If the chromatographic data is acquired and processed by a computer, then with modern software, a separation of 4σ is quite adequate.

Calculating the Efficiency Required to Achieve a Specific Resolution

Having defined that the resolution required to separate the critical pair in a specific sample is 4σ it is now possible to calculate the number of theoretical plates that are necessary to provide adequate quantitative accuracy. This can be easily carried out using the information provided by the Plate Theory in the chapter 2. Restating figure 10 from chapter 2 as figure 8, it is seen that the retention volume difference between the peaks (Δv) is

$$\Delta V = n(K_B - K_A)v_v$$

and bearing in mind $\alpha = \dfrac{K_B}{K_A}$, $\quad \Delta V = nK_A(\alpha - 1)v_s \qquad (12)$

By definition (Δv) must be equivalent to (4σ) and from figure 8,

$$\Delta V = 4\sqrt{n}(v_m + K_A vs) \qquad (13)$$

Equating equations (12) and (13),

$$4\sqrt{n}\,(v_m + K_A v_s) = nK_A(\alpha - 1)v_s$$

Figure 8

A Chromatogram Showing the Separation of Two Solutes

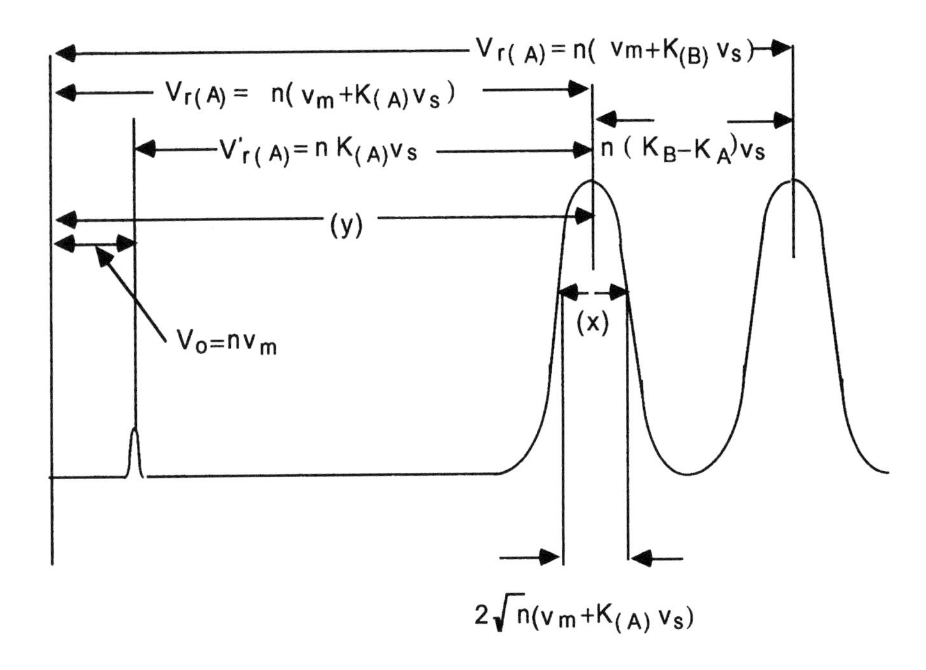

Dividing throughout by (v_m) and remembering $k'_A = \dfrac{K_A v_s}{v_m}$

$$4\sqrt{n}\,(1 + k'_A) = n\,k'_A(\alpha - 1)$$

Solving for (n), $$n = \left(\frac{4(1 + k'_A)}{k'_A(\alpha - 1)}\right)^2 \qquad (14)$$

Equation (14) was first developed by Purnell in 1959 (7) and has proved to be one of the most important equations in column design and one that is the greatest use as an aid in column selection for the practicing analyst. Employing the data from the reduced chromatogram, equation (14) will allow the analyst to calculate the number of theoretical plates that are necessary to separate the sample into its constituents in a manner that will provide an accurate quantitative analysis. In fact, it will also allow the optimum column to be designed that will provide the analysis in the minimum time, given the equipment available (1). However, this would only be advisable if the analysis was to be routine with many samples being analyzed every day. In general, calculating the required efficiency using equation (14) allows the analyst to select the most appropriate column available, that has the required efficiency or perhaps a little in excess.

Peak Asymmetry

Although most concentration profiles of peaks eluted from a liquid chromatograph are symmetrical in shape and closely approximate to a Gaussian curve, under some circumstances peak asymmetry will occur. There are a number of reasons for peak asymmetry. One cause results from the non equivalence of the resistance to mass transfer terms in the HETP equation. Because there is always a net transfer of solute from the mobile phase to the stationary phase in the front of the peak, the resistance to mass transfer dispersion effect in the mobile phase will dominate slightly over the dispersion due to the resistance to mass transfer in the stationary phase in its contribution to the overall variance of the peak. Now, the mass transfer in the mobile phase is enhanced by the mixing that results from the eddies caused by the continual change in direction of the mobile phase as it flows between the particles of the packing. Consequently, the resistance to mass transfer in the mobile phase may become significantly less than that in the stationary phase. As a result, the variance of the front half of the peak may be less than the variance of the latter half of the peak and thus, peak asymmetry will result. Another cause of peak asymmetry is the heat evolved and adsorbed during the distribution process. As solute is absorbed into the stationary phase, the heat of adsorption is evolved and the local temperature rises reducing the magnitude of the distribution coefficient in the front half of the peak.

As the peak velocity is inversely proportional to the magnitude of the distribution coefficient, this results in the front part of the peak moving through the column slightly faster than the latter half of the peak. Conversely, as solute is desorbed from the stationary phase into the mobile phase, the heat of adsorption is absorbed and the local temperature falls, increasing the magnitude of the distribution coefficient in the rear half of the peak. This results in the rear part of the peak moving through the column slightly slower than the front half of the peak. These combined effects result in peak asymmetry.

However, the major contribution to peak asymmetry is usually a result of column overload and the two effects that can occur are depicted in figure 9.

Figure 9

Peak Asymmetry Resulting from Column Overload

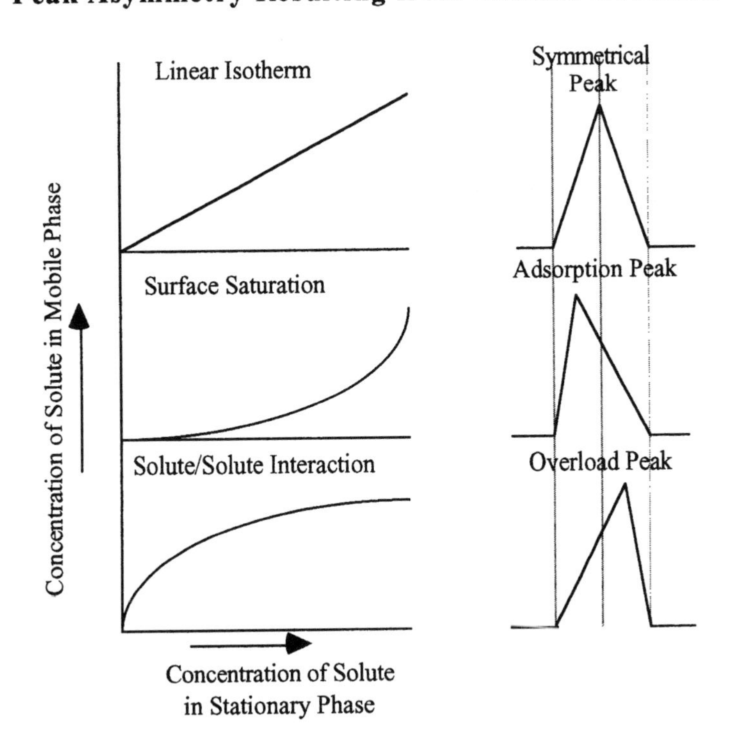

The upper curve shows the adsorption isotherm that normally occurs in liquid chromatography separations where the concentration of solute in the system is very low. The isotherm is linear and thus the distribution coefficient is constant at all concentrations of solute in either phase. It follows that as the peak velocity is inversely related to the distribution coefficient, all solute concentrations travel at the same velocity through the column and the peak is symmetrical.

The second isotherm shown in figure 9 can result from surface saturation of the stationary phase. If the stationary phase has a limited surface area and the solute molecules are large, then even at low concentrations of solute a significant proportion of the surface is covered. As a result, the probability of a solute molecule interacting with the surface will be reduced at higher solute concentrations. This means that the adsorption isotherm takes the form shown in the middle section of the diagram. It also follows that at the higher concentrations, the distribution coefficient of the solute with respect to the stationary phase is reduced. Consequently, the higher concentrations of the solute move more rapidly through the column than the lower concentrations resulting in the type of asymmetry termed an adsorption peak. This term for this type of asymmetry is not ideal and arises from the fact that this shaped peak commonly occurred in gas-solid chromatography. For the want of a better term, it has persisted and is used to describe a peak were the peak maximum is displaced toward the front of the peak.

The lower isotherm represents the overload condition that can occur in liquid/liquid or gas/liquid systems under somewhat unique circumstances. If the interactions between solute molecules with themselves is stronger than the interactions between the solute molecules and the stationary phase molecules, then, as the concentration of solute molecules increases, the distribution coefficient of the solute with respect to the stationary phase also increases. This is because the solute molecules interact more strongly with a solution of themselves in the stationary phase than the stationary phase alone. Thus, the higher concentrations of solute in the chromatographic system will move more slowly through the system than the lower concentrations which results in a peak shape shown as an overload peak. This type of peak, for obvious reasons, is frequently found in gas-liquid chromatography (from whence it received its name) but infrequently in liquid-solid chromatography.

Column Selection

Unfortunately, although there are a number of different stationary phases available, there is only a very limited range of column sizes. If a column of specific dimensions is needed for a fully optimized column system, then it will need to be packed by the analyst or a supporting group. Very few manufacturers will custom-pack a specific column and if they do, it is likely to be very costly. On the other hand, packing columns is a tedious procedure and requires some relatively expensive apparatus. Furthermore, it is necessary to pack columns at pressures often exceeding 12,000 p.s.i. which demands special high pressure fittings and appropriate safety precautions.

For the average analytical laboratory it is advisable to avoid packing columns if at all possible. For a large analytical service group handling hundreds or even thousands of sample a day, it may be economical to design and pack optimized columns for specific, high-throughput LC analyses. Under such circumstances a laboratory fitted out for column packing might be economically advantageous.

In most cases the analyst must choose a column size that is commercially available and fortunately most manufactures provide an appropriate range of sizes. It is interesting to note that, although the predictions of Martin and Synge were correct and small particles give the smallest HETP and the highest efficiencies, there is a caveat to this argument.

The tentative assumption behind this statement is that there is always sufficient pressure available to provide the necessary flow-rate. In practice this is not the case. LC pumps have a maximum operating limit which is generally about 6,000 or 10,000 p.s.i. but, unfortunately, it is not the pump that usually controls the limit of the operating pressure. It is the sample valve that usually limits the pressure and although sample valves are rated to operate at pressures up to 10,000 p.s.i. they cannot do so over long periods of time particularly if automatically actuated. This is because the spigot and seating are compressed at high pressure to prevent leaks. When solid contaminants that are of microscopic size and quantity come in contact with the valve seat, it rapidly wears and begins to leak. Unfortunately, such contaminants are almost impossible to eliminate.

In practice, for the long-term satisfactory use of a sample valve specified for operation at 6,000 or even 10,000 p.s.i., it should not be subjected to hydraulic pressures in excess of 3,000–3,500 p.s.i. under continuous operation. It follows that if the pressure is limited to a maximum, albeit by the operational pressure limit of the pump or by the limiting pressure for the long term use of the sample valve, there will be a limit also placed on the particle size that can be used for a column of given length. This is because the column packing must be sufficiently permeable so that the mobile phase velocity at the maximum pressure will be at least as great as the optimum as defined by the HETP equation. If a higher efficiency is required, the column must be made longer, then, if the same particle size is employed, the permeability will be inadequate and the maximum permissible pressure will not be sufficient to provide the optimum mobile phase velocity. Under such circumstances the particle size must be made larger.

If the operating pressure of a column is limited, simple separations will be achieved very rapidly on short columns packed with very ***small*** *particles. In contrast, difficult separations will take longer and will require long columns packed with* ***larger*** *particles.*

The family of columns shown below in table 1 is suggested for general use which, with the exception of the column 148 cm long, are mostly commercially available.

Table 1 Design Data on a Series of Optimized Columns

Separation Ratio	Particle Diameter (micron)	Column Length (cm)	Analysis Time	Peak Capacity
1.02	**10.5**	**148**	**3.6 hr**	**115**
1.03	**7.7**	**58.4**	**58.4 min**	**84**
1.050	**4.6**	**12.6**	**8.2 min**	**51**
1.080	**2.9**	**3.1**	**74.7 sec**	**32**

The above columns were calculated for a sample where the first peak of the critical pair in the reduced chromatogram was eluted at a (k') of 2.5 and the last peak was eluted at a (k') of 5.0. The method of calculation is given in (1).

It is seen that, although the dimensions and particle sizes may not be precisely matched, all three columns are of a size closely similar to those commercially available with, perhaps, the exception of the long high efficiency column. The small 3 cm column is excellent for the preliminary assessment of a sample. As a result of its size it does not use large volumes of solvent and can be quickly reconditioned after a separation in readiness for the next run. It is very convenient for choosing the best phase system in method development. The other columns would be chosen on a basis of the efficiency required to separate the critical pair in the reduced chromatogram of the sample for analysis.

Column Diameter

Little has been said concerning the column diameter which, unfortunately, is an aspect of column technology that involves extensive theoretical discussion which is probably not appropriate here. Each column that is optimized to analyze a particular sample in the minimum time and with the minimum solvent consumption will also have an *optimum diameter*. The optimum column diameter provides the chromatographic system with the *maximum mass sensitivity* which is sometimes necessary when there is limited sample available. In general, short columns packed with small particles should have a relatively large diameter. Long columns packed with larger particles will have small diameters. For example the column 3 cm long in the above table should be about 4 mm in diameter, the column 12.5 cm long about 3 mm in diameter whereas the 150 cm long column would probably be about 1 mm in diameter. The precise optimum column diameter will depend on the type of sample and the specifications of the chromatograph, but the above values will be fair approximation for most practical purposes.

Preparative Columns

Preparative chromatography, as opposed to analytical chromatography, involves the collection of a particular component of the sample for further examination or processing. The scale of the separation can extend from a few micrograms of pure solute, collected for subsequent spectroscopic examination, to a kilo or more of the material separated for production purposes on a column, perhaps a meter or more in diameter. The analyst, however, is involved very rarely in large scale preparative separations where the amount of material collected is much greater than a few milligrams. Preparative procedures are sometimes required to enable the analyst to obtain material for structural elucidation, confirm compound identity for forensic purposes or, in some cases, to assess its physiological activity.

The easiest way for an analyst to obtain small quantities of a component of a mixture is to *overload* an *analytical column*. In order to exercise this technique, the solute of interest must be well separated from its closest neighbor. The column can then be overloaded with sample until the peak dispersion resulting from the overload, causes the two peaks to touch at their base. There are two types of column overload, *volume overload* and *mass overload*. In practice, it is often advantageous to employ a combination of both methods and a simple procedure for doing this will be given overleaf.

Volume overload can be treated in a simple way by the plate theory (8,9). In contrast, the theory of mass overload is complicated (10-12) and requires a considerable amount of basic physical chemical data, such as the adsorption isotherms of the solutes, before it can be applied to a practical problem. Volume overload is useful where the solutes of interest are relatively insoluble in the mobile phase and thus, to apply a sample of sufficient size onto the column, a large sample volume is necessary. If the sample is very soluble in the mobile phase then mass overload might be appropriate.

Volume Overload

In figure 10 the solute of interest and its closest neighbor are depicted.

Figure 10

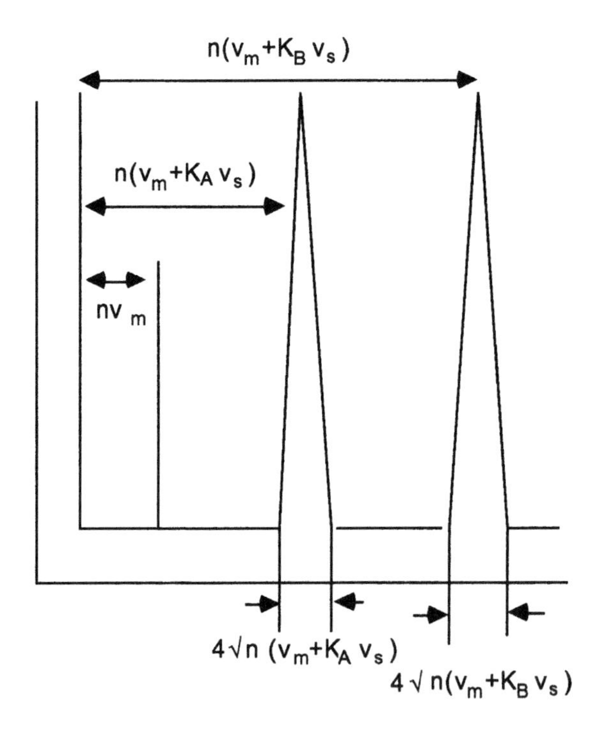

To determine the band dispersion that results from a significant, but moderate, sample volume overload the summation of variances can be used. However, when the sample volume becomes excessive, the band dispersion that results becomes equivalent to the sample volume itself. In figure 10, two solutes are depicted that are eluted from a column under conditions of no overload. If the dispersion from the excessive sample volume just allows the peaks to touch at the base, the peak separation in milliliters of mobile phase passed through the column will be equivalent to the sample volume (V_i) plus half the base width of both peaks. It is assumed in figure 10 that the efficiency of each peak is the same and in most cases this will be true. If there is some significant difference, an average value of the efficiencies of the two peaks can be taken.

It is seen that if (V_i) is the injection volume, then

$$n(v_m + K_B v_s) - n(v_m + K_A v_s) = V_i + 2\sqrt{n}(v_m + K_B v_s) + 2\sqrt{n}(v_m + K_A v_s)$$

Now as $$V_r = n(v_m + K v_s)$$

Then, $$V_{r(B)} - V_{r(A)} = V_i + \frac{2(V_{r(B)} + V_{r(A)})}{\sqrt{n}}$$

Rearranging,

$$V_i = (V_{r(B)} - V_{r(A)}) - \frac{2(V_{r(B)} + V_{r(A)})}{\sqrt{n}} \qquad (15)$$

Equation (15) allows the analyst to calculate the sample volume that can be used before loss of resolution from a knowledge of the retention volume of the solute of interest and its nearest neighbor and the efficiency of the column.

An even simpler approximation would be

$$V_i = (V_{r(B)} - V_{r(A)}) \qquad (16)$$

Volume overload employing a solution of the material in the mobile phase at a level of about 5% w/v is a recommended method of sampling for preparative columns if the system is not optimized. However, a combination of volume overload and mass overload has also been suggested as an alternative procedure by Knox (13).

Knox and Piper (13) assumed that the majority of the adsorption isotherms were, indeed, Langmuir in form and then postulated that all the peaks that were 'mass overloaded' would be approximately triangular in shape. As a consequence, Knox and Piper proposed that mass overload could be treated in a similar manner to volume overload. Whether all solute/stationary phase isotherms are Langmuir in type is a moot point and the assumption should be taken with some caution. Knox and Piper then suggested that the best compromise was to utilize about half the maximum sample volume as defined by equation (15), which would then reduce the distance between the peaks by half. They then recommended that the concentration of the solute should be increased until dispersion due to mass overload just caused the two peaks to touch.

Knox summarized his recommendations in the following way (14).

1/ Develop an analytical separation which gives the best possible resolution between the critical solutes.

2/ Determine the difference between the retention volumes of the two solutes (ΔV).

3/ Employ sample volumes of about 0.4 DV and increase the concentration of solute until the gap between the pair is just filled.

Unfortunately, this procedure can only be successful if the critical pair can be well resolved and column overload is a practical solution to the problem. Often, due to the complex nature of practical mixtures, values for these conditions are not realized and the optimum *column* may need to be designed (1) which, in the design procedure, will automatically be given a radius that will accommodate the load that is required. Column design and construction, however, are not usually considered part of the responsibilities of the analyst.

References

1/ *"Liquid Chromatography Column Theory"*, R.P.W. Scott, John Wiley and Sons, Chichester-New York-Singapore, (1992).

2/ J. C. Giddings, *J. Chromatogr.*, **5**(1961)46.

3/ J. F. K. Huber and J. A. R. J. Hulsman, *Anal. Chim ., Acta.*, **38**(1967)305.

4/ G. J. Kennedy and J. H. Knox, *J. Chromatogr. Sci.*, **10**(1972)606.

5/ Cs. Horvath and H.J. Lin, *J. Chromatogr.*, **149**(1976)401.

6/ J.J. Van Deemter, F.J. Zuiderwg and A. Klinkenberg, *Chem. Eng. Sci.*, **5**(1956)271.

7/ J.H. Purnell, *Nature,* **No. 4704, Dec. 9**(1959)2009.

8/ R.P. W. Scott and P. Kucera, *J. Chromatogr.*,**119**(1976)467.

9/ J.H. Knox and M. Saleen, *J. Chromatogr.Sci.*.**7**(1969)614.

10/ S. Golshan-Shirazi, A. Jaulmes and G. Guiochon, *Anal. Chem.,* **60**(1988)1856.

11/ S. Golshan-Shirazi,A. Jaulmes and G. Guiochon, *Anal. Chem.,* **61**(1989)1276.

12/ S. Golshan-Shirazi, A. Jaulmes and G. Guiochon, *Anal. Chem.,* **61**(1989)1368.

13/ J.H. Knox and H.P. Piper, *J. Chromatogr.* **363**(1986)1.

14/ J.H. Knox. *The Chromatographic Society Bulletin,* **29**(1988)18.

5
The Liquid Chromatograph

The original LC chromatographic system of Tswett consisted of a vertical glass tube, a few centimeters in diameter and about 30 cm high, packed about half way up with the adsorbent (calcium carbonate). The extract of plant pigments was placed on the top of the packing and the mobile phase carefully added to fill the tube. The solvent percolated through the packing under gravity, developing the separation which was observed as it proceeded by the different colored bands at the wall of the tube. The simple apparatus of Tswett contained all the essentials of a modern chromatograph but, although he might have recognized the modern LC column for what it is, it is doubtful if the modern liquid chromatograph would have had any meaning for him.

The contemporary chromatograph used for analytical purposes is a very complex instrument that may operate at pressures up to 10,000 p.s.i. and provide flow rates that range from a few microliters per minute to 10 or 20 ml/minute. Solutes can be detected easily at concentration levels as low as $1x10^{-9}$ g/ml and a complete analysis can be carried out on a few micrograms of sample in a few minutes. The range of liquid chromatographs that is available extends from the relatively simple and inexpensive instrument, suitable for the majority of routine analyses, to the very elaborate and expensive machines that are more appropriate for analytical method development.

The Basic Liquid Chromatograph

The basic liquid chromatograph may be considered to be comprised of six units. The mobile phase storage and supply system, the pump and programmer, the sample valve, the column, the detector and finally a means of presenting the chromatogram and possibly calculating the results. A block diagram of the basic liquid chromatograph is shown in figure 1.

Figure 1 The Basic Liquid Chromatograph

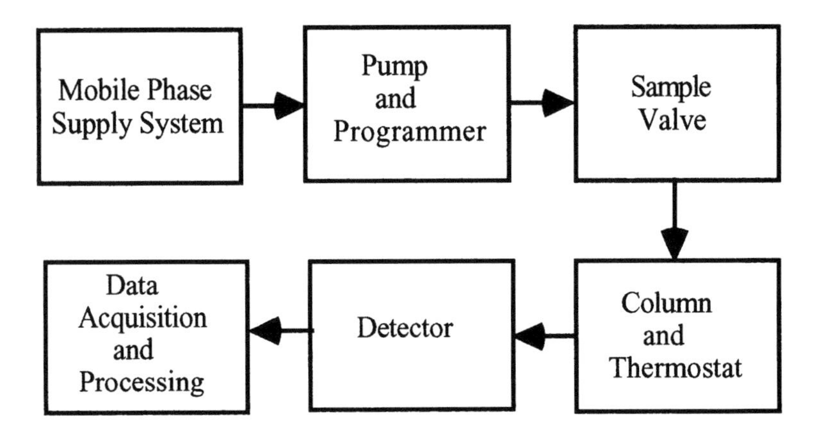

The components of the basic chromatograph will now be individually discussed.

The Mobile Phase Supply System

The mobile phase supply system consists of a series of reservoirs normally having a capacity ranging from 200 ml to 1,000 ml. Two reservoirs are the minimum required and are usually constructed of glass and fitted with an exit port open to air. Stainless steel is an alternative material for reservoir construction but is not considered satisfactory for mobile phases buffered to a low pH and containing certain materials that can cause erosion of the stainless steel. Each reservoir is usually fitted with a gas diffuser through which helium can be bubbled. Many solvents and solvent mixtures, particularly *aqueous* mixtures, contain significant amounts of dissolved nitrogen and oxygen from the air. These gasses form bubbles in the chromatographic system that can cause serious detector noise and loss of column efficiency. Helium is very insoluble in solvents, and thus it can purge the oxygen and nitrogen from the solvent but does not dissolve itself and produce bubbles in the system. Alternatively, a vacuum can be applied to the reservoir to remove dissolved air but this is not a permanent solution as on releasing the vacuum to allow the solvent to pass to the pump, air again dissolves in the solvent. The solvent is drawn from the reservoir through a stainless steel or sintered glass filter to remove any solid contaminants and direct to the pump or mixer. Depending on the type of programmer employed, the supply from each reservoir may pass to a pump or to the blending device of the programmer. In general, the solvent reservoirs are not thermostatted but, if necessary, the solvent or solvent mixture is brought to the column temperature by an appropriate heat exchanger situated in the column thermostat. The solvent containers are often situated in an enclosure to avoid moisture vapor being absorbed from the air. Such containers also protect the user from toxic solvent vapors such as chloroform or aromatic hydrocarbons should they be employed as a component of the mobile phase.

The Gradient Programmer

There are two types of gradient programmer. In the first type, the solvent mixing occurs at high pressure and in the second the solvents are premixed at low pressure and then passed to the pump. The high pressure programmer is the simplest but most expensive as it requires a pump for each solvent supply. There can be any number of solvents involved in a mobile phase program, however, the majority of LC analyses usually require only two solvents but up to four solvents can be accommodated. The layout of a high pressure gradient system is shown in figure 2 and includes, as an example, provision for three solvents to be mixed by appropriate programming.

Figure 2

The High Pressure Gradient Programmer

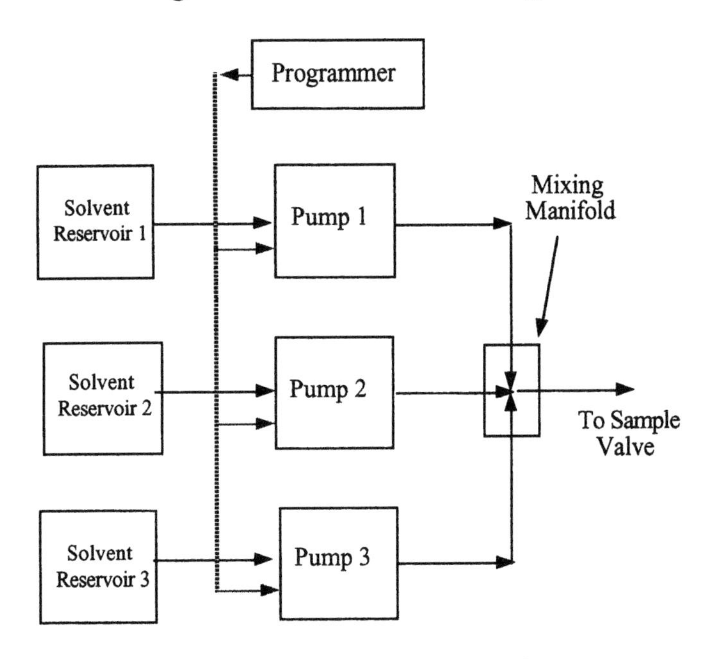

Each solvent passes from its reservoir directly to a pump and from the pump to a mixing manifold. After mixing, the solvents pass to the sample valve and column. The pumps control the actual program and are usually driven by stepping motors. The volume delivery of each solvent is controlled by the speed of the respective pump. In turn, the speed of each motor is precisely determined by the frequency of its power supply which can be either generated by external oscillators or, if the chromatograph is computer controlled, directly from the computer itself.

In a low pressure programmer, the solvent from each reservoir passes to an oscillating valve, instead of flowing directly to a pump, the output from which is connected to a mixing manifold. A diagram of the layout of a typical low pressure solvent programmer is shown in figure 3.

Figure 3

A Low Pressure Solvent Programmer

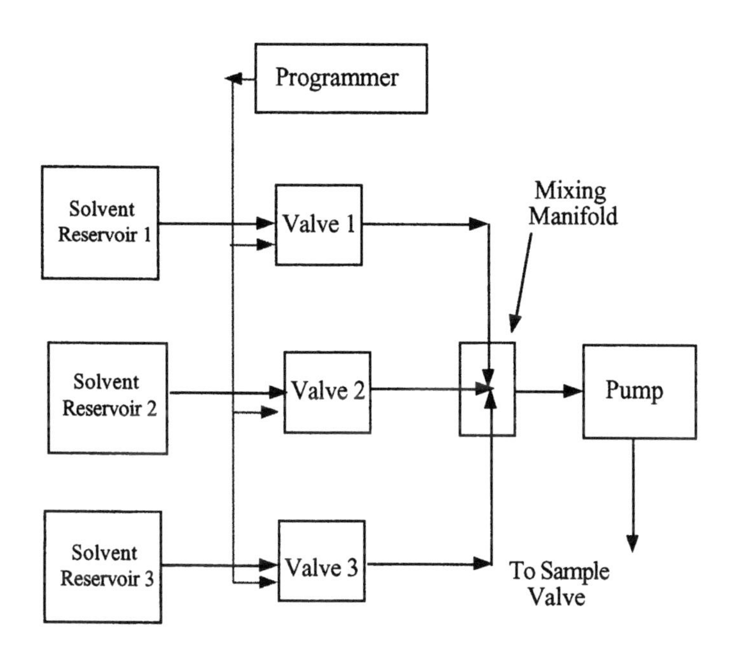

The manifold also receives and mixes solvents from each of the programmed valves. The valves are electrically operated and programmed to open and close for different periods of time by adjusting the frequency and wave form of the supply. Thus, a predetermined amount of each solvent is allowed to flow into the manifold. The valves can be driven by oscillators contained in a separate electronic programmer or if the chromatograph is computer controlled, the controlling waveform and frequency can be provided directly from the computer.

The LC Pump

There are a number of different types of pumps that can provide the necessary pressures and flow rates required by the modern liquid chromatograph. In the early years of the LC renaissance, there were two types of pump in common use; they were the pneumatic pump, where the necessary high pressures were achieved by pneumatic amplification, and the syringe pump, which was simply a large, strongly constructed syringe with a plunger that was driven by a motor. Today the majority of modern chromatographs are fitted with reciprocating pumps fitted with either pistons or diaphragms. An example of each type of pump will now be described.

The Pneumatic Pump

The pneumatic pump has a relatively large flow capacity but, today, is largely used for column packing and not for LC analyses. It can provide extremely high pressures and is relatively inexpensive, but the high pressure models are a little bulky. A diagram of a pneumatic pump is shown in figure 4.

Figure 4

A Diagram of the Pneumatic Pump

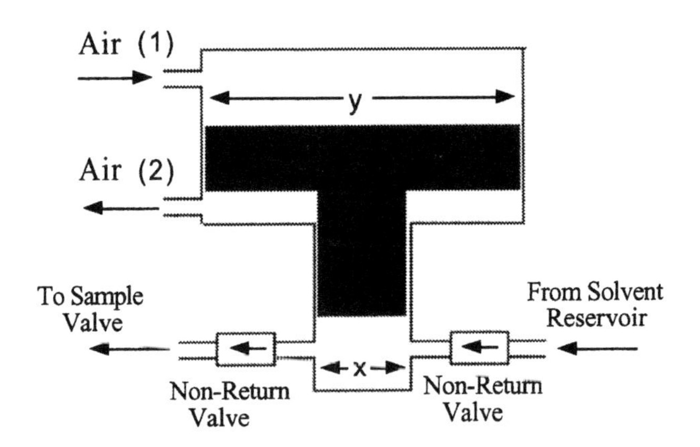

It is seen in figure (4) that the total air pressure on the piston, diameter (y), is transferred to a piston controlling the liquid pressure of diameter (x). Because the radii of the pistons differ, there will be a net pressure amplification of $\frac{y^2}{x^2}$. If, for example, the upper piston is 10 cm in diameter and the lower piston 1 cm in diameter, then the amplification factor will be $\frac{10^2}{1^2} = 100$. It is seen that the system can provide very high pressures in a relatively simple manner. This type of pump is usually employed for column packing. The operation of the pump is as follows. Air enters port (1) and applies a pressure to the upper piston that is directly transmitted to the lower piston. If connected to a liquid chromatograph, or column packing manifold, liquid will flow out of the left hand side non-return valve as shown in the diagram. This will continue until the maximum movement of the piston is reached. A micro switch is then activated and the air pressure transferred to port (2). The piston moves upward drawing mobile phase through the right hand non-return valve filling the cylinder with solvent. When the piston reaches the limit of a second micro switch the air supply is returned to port (1) and the process repeated. The refilling process is extremely fast and, if an appropriate pulse dampener is used, the outlet pressure remains reasonably constant during the refill cycle.

All LC pumps require the use of a non-return valve and so the function of non-return valves will now be described.

Non-Return Valves

Non-return valves are essential in the operation of pumps. It is important to ensure that when the pump compresses the solvent, it leaves at the exit port and is completely arrested at the inlet port. Conversely when the pump draws fresh solvent into the cylinder, the non-return valves must allow solvent to flow through the inlet valve, but flow-back from the other side of the exit valve must be completely arrested. Most non-return valves are of similar design and the construction of a typical non-return valve is shown in figure 5.

Figure 5
The Design of a Typical Non-Return Valve

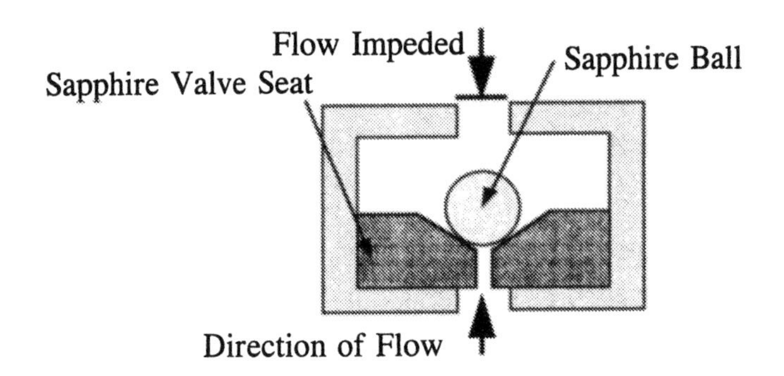

The critical part of the valve consists of a synthetic sapphire ball resting on a seat. The seat may be of stainless steel, PTFE or, more usually, also of sapphire. When the flow is directed against the ball the ball moves forward allowing the liquid to flow past it. When the direction of pressure changes resulting in potential flow-back through the valve, the ball falls back on its seat and arrests the flow.

Figure 6
Twin Non-Return Valve Combination

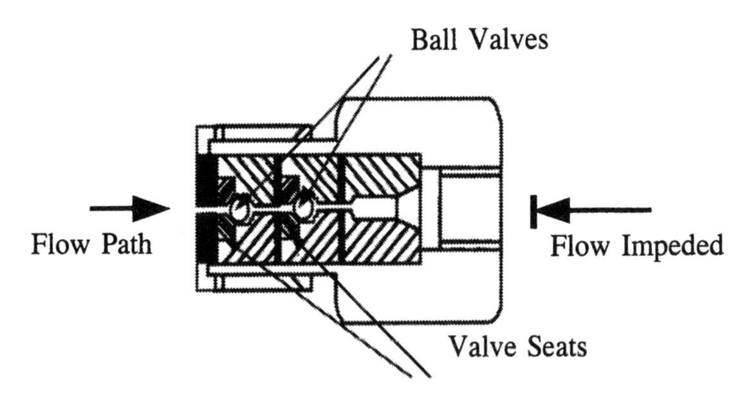

Courtesy of HITACHI Instruments Inc.

With careful design and exacting construction these types of valves can be extremely efficient. In practice, to ensure absolute minimum back flow through the valve, a single non-return valve assembly usually contains *two* non-return ball valves connected in series as shown in figure 6.

The Syringe Pump

The syringe pump, as the name implies, is a large, electrically operated simulation of the hypodermic syringe. Although used in the early days during the renaissance of LC, it is rarely used today as, due to its design, it can provide only a limited pressure and the volume of mobile phase available is restricted to the pump volume. Unless the separation is stopped while the pump is refilled and the development subsequently continued, the pump can only develop a separation to the limit of its own capacity. A diagram of a syringe pump is shown in figure 7.

Figure 7

The Syringe Pump

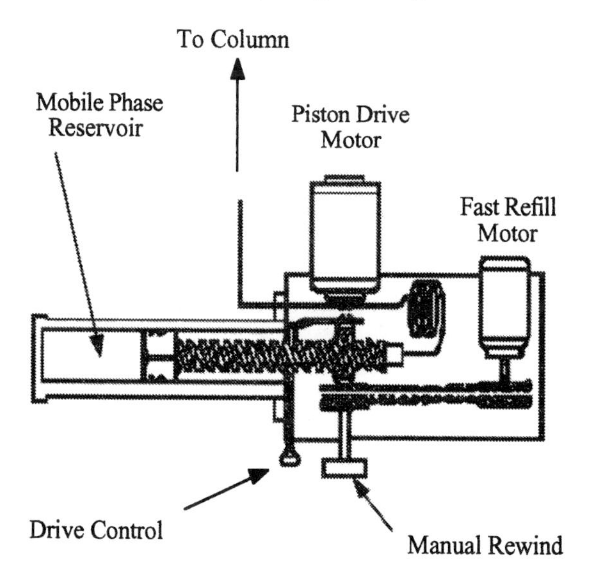

Courtesy of the Perkin Elmer Corporation

The pump consists of a large metal syringe, the piston being propelled by an electric motor and driven by a worm gear. The speed of the motor determines the pump delivery. Another motor actuates the piston by a different system of gearing to refill the syringe rapidly when required. The solvent is sucked into the cylinder through a hole in the center of the piston and between the piston and the outlet there is a coil that acts as a dampener. This type of pump is still occasionally used for the mobile phase supply to microbore columns that require small volumes of mobile phase to develop the separation. It is also sometimes used for reagent delivery in post column derivatization as it can be made to deliver a very constant reagent supply at very low flow rates.

The Single Piston Reciprocating Pump

The single piston reciprocating pump was the first of its type to be used in liquid chromatographs and is still very popular today.

Figure 8

A Single Piston Reciprocating Pump

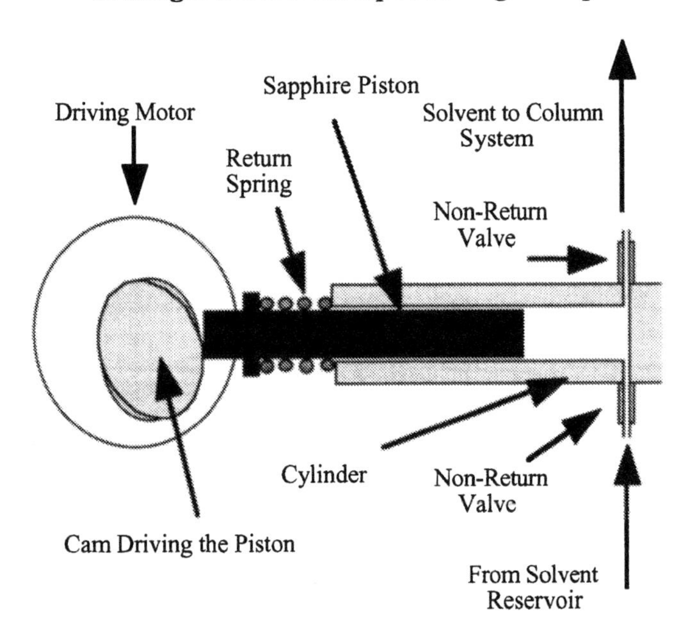

It is relatively inexpensive and allows the analyst to become involved in liquid chromatography techniques without excessive investment. A diagram of the single piston pump is shown in figure 8. The piston is made of synthetic sapphire, in fact most pistons of modern LC pumps are made of sapphire to reduce wear and extend the working life of the pump. The cylinder is usually made of stainless steel which incorporates two non-return valves in line with the inlet and outlet connections to the pump. The piston is driven by a stainless steel cam which forces the piston into the cylinder expressing the solvent through the exit non-return valve. After reaching the maximum movement, the piston follows the cam and returns as a result of the pressure exerted by the return spring. During this movement the cylinder is loaded with more solvent through the inlet non-return valve. The shape of the cam is cut to provide a linear movement of the piston during expression of the solvent but a sudden return movement on the refill stroke. In this way the pulse effect is reduced. However, the pulses are not completely eliminated and the presence of these pulses in the flow of mobile phase is probably the most serious disadvantage of the single piston pump. Nevertheless, as a result of its low cost it remains one of the more popular LC pumps.

The Rapid Refill Pump

In order to avoid the pulses resulting from a single piston refill pump, rapid refilling systems have been devised. These have varied between cleverly designed actuating cams to drive the piston to electronically operated piston movement. An interesting and successful approach to this problem is exemplified by the pump design shown in figure 9.

The pump consists of two cylinders and a single common piston. The expression of the solvent to the column is depicted in the upper part of figure 9. As the piston progresses to the right, solvent is pumped to the column system but at the same time fresh solvent is being withdrawn into the right hand chamber from the solvent supply system. At the point where the piston arrives at the extent of its travel, a step in the driving cam is reached and the piston is rapidly reversed.

As a result the contents of the chamber on the right-hand side are conveyed to the left-hand chamber. This situation is depicted in the lower part of figure 9. The transfer rate of the solvent to the left-hand chamber is 100 times as fast as the delivery rate to the column and consequently reduces the pulse on refill very significantly. Furthermore, if a solvent gradient is being used and the right-hand chamber is being filled with a solvent mixture, excellent mixing is achieved during the refill of the left-hand chamber.

Figure 9

The Rapid Refill Pump

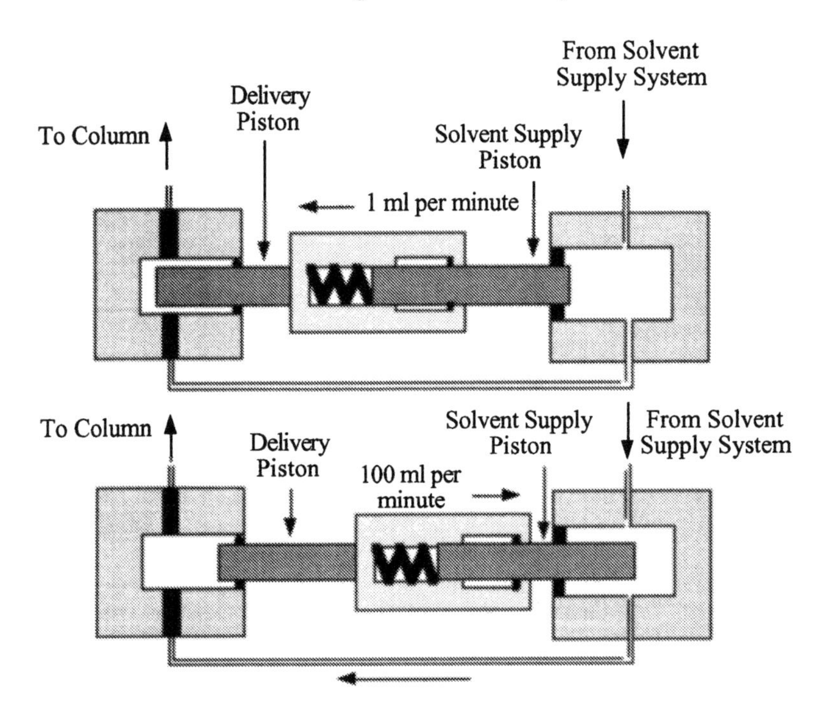

Courtesy of Perkin Elmer Inc.

An alternative approach to the removal of pump pulses, which is probably the more successful but, as one might expect, the more expensive, is the use of twin pump heads. In a two-headed pump, one cylinder can be filled while the other is delivering solvent to the column.

The Twin-Headed Pump

The actual cylinders and pistons of a two-headed pump are constructed in a very similar manner to the single piston pump with a sapphire piston and a stainless steel cylinder. Each cylinder is fitted with non-return valves both at the inlet and outlet. The cams that drive the two pistons are carefully cut to provide an increase in flow from one pump while the other pump is being filled to compensate for the loss of delivery during the refill process and thus, a fall in pressure. A diagram of a twin-headed pump is shown in figure 10.

Figure 10

The Twin-Headed Pump

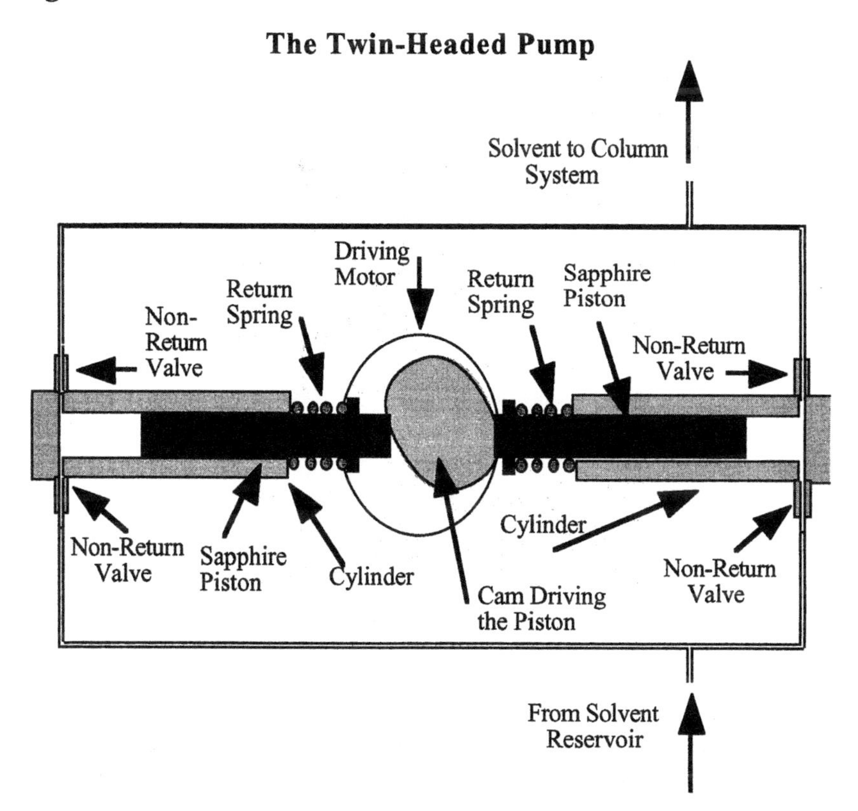

It is seen that there is a common supply of mobile phase from the solvent reservoir or solvent programmer to both pumps and the output of each pump joins and the solvent then passes to the sample valve and then to the column. In the diagram, a single cam drives both pistons, but in practice, to minimize pressure pulses, each pump usually has its own cam drive from the motor. The cams are carefully designed to produce a virtually pulse-free flow. The displacement volume of each pump can vary from 20 or 30 microliters to over one milliliter but the usual displacement volume for the Waters pump is about 250 ml. The pump is driven by a stepping motor and thus the delivery depends on the frequency of the supply fed to the motor. This gives the pump a very wide range of delivery volumes to choose from, extending from a few microliters per minute to over 10 ml per minute. The form of the Waters twin headed pump is shown in figure 11.

Figure 11

The Waters 501 Twin-Headed LC Pump

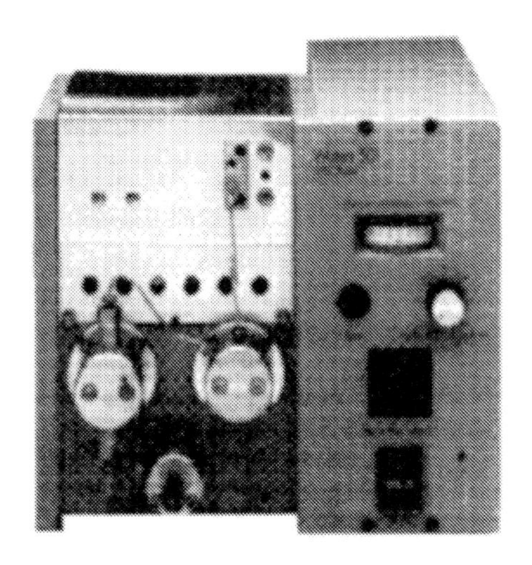

Courtesy of Perkin Elmer Inc.

The Waters pump was one of the first twin headed LC pumps to be produced and was probably the major factor in establishing Waters fine reputation in the field of liquid chromatography instrumentation.

The Diaphragm Pump

In the reciprocating diaphragm pump the actuating piston does not come into direct contact with the mobile phase and thus the demands on the piston-cylinder seal are not so great. Because the diaphragm has a relatively high surface area, the movement of the diaphragm is relatively small and thus the pump can be operated at a fairly high frequency. High frequency pumping results in a reduction in pulse amplitude. Furthermore, high frequency pulses are more readily damped by the column system. It must be emphasized, however, that diaphragm pumps are not pulseless. A diagram of a diaphragm pump, showing its mode of action, is depicted in figure 12.

Figure 12

The Action of a Diaphragm Pump

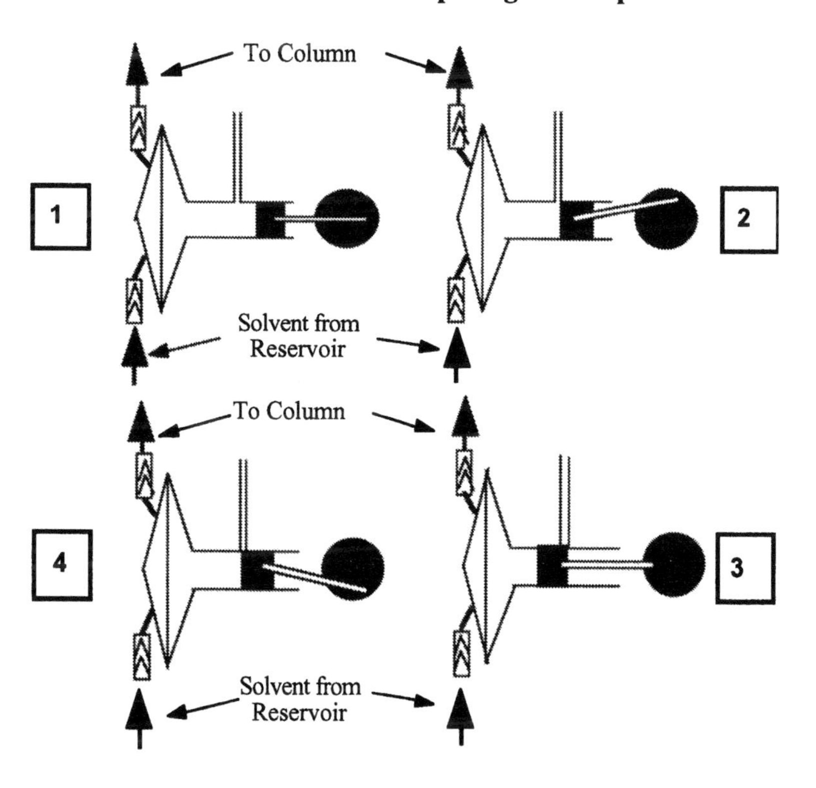

The wheel driving the crank is rotating anti-clockwise and in position (1) the diaphragm has been withdrawn and the pumping cavity behind the diaphragm filled with solvent. In stage (2), the piston advances and when it passes the pumping fluid inlet, it starts compressing the diaphragm expressing solvent to the column. In stage (3) the diaphragm has been compressed to its limit and the piston starts to return. In stage (4) the piston moves back withdrawing the diaphragm sucking liquid into the pumping cavity ready for the next thrust. The inlet from the mobile phase supply and the outlet to the column are both fitted with non-return valves in the usual manner. Due to the relatively large volume of the pumping cavity, this type of pump is not often used in analytical chromatographs but is frequently employed for preparative chromatography.

The Sample Valve

There are basically two types of LC sample valve, those with an internal loop and those with an external loop. Valves with an internal loop are normally designed to deliver sample volumes of less than one microliters. Valves with external loops can deliver sample volumes ranging from a few microliters to several milliliters or more. In general, LC sample valves must be able to sustain pressures up to 10,000 p.s.i., although they are likely to operate on a continuous basis, at pressures of 3,000 p.s.i. or less.

The effect of fluid pressure on the sample valve, from the point of view of its impact on its operating lifetime, has already been touched upon. The higher the operating pressure the tighter the valve seating surfaces must be forced together to eliminate any leak. It follows that any abrasive material, however fine, that passes into the valve can cause the valve seating to become seriously scored each time it is rotated which will ultimately lead to leaks. This will mean that the sample size will start to vary between samples and eventually seriously affect the accuracy of the analysis. It follows that any solid material must be carefully removed from any sample before filling the valve. (The subject of solid elimination from the sample will be discussed under sample preparation. However, even with the most stringent precautions taken, all solid material cannot be completely removed from a sample. Furthermore, at the column side of the sample valve, if silica based packings are being employed, highly abrasive material that has a particle size less than the porosity of the column frit can swirl back into the valve on injection also scoring the valve seat. Unfortunately, the only solution to this problem is to avoid, as far as is possible, operating the valve and column system at excessively high pressures. It has been suggested in an earlier chapter that a good compromise between ensuring satisfactory column efficiency and resolution by the use of adequate pressure and at the same time to achieve reasonable valve life is to endeavor *not* to operate the system at pressures much *above* 3,000 p.s.i.

As a consequence of the high pressures that must be tolerated, LC sample valves are usually made from stainless steel. The exception to the use of stainless steel will arise in biochemical applications where the materials of construction may need to be bio-compatible. In such cases the valves may be made from titanium or some other appropriate bio-compatible material.

Figure 13
A Simple Form of the LC Sample Valve

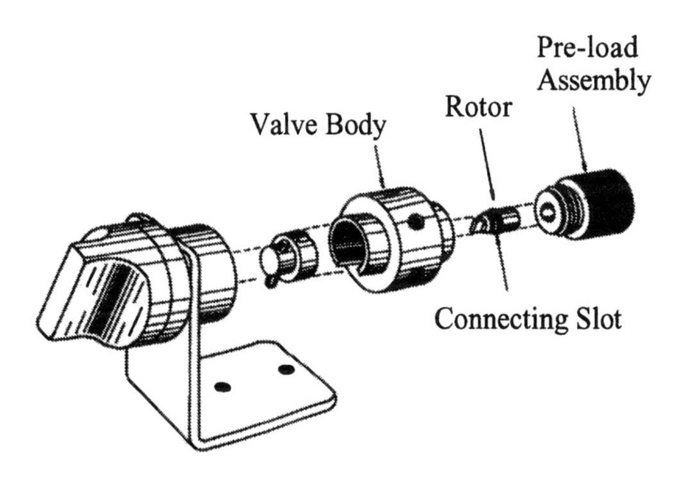

Courtesy of Valco Instruments Inc.

It should be stressed that only those surfaces that actually come in contact with the sample need to be bio-compatible and the major parts of the valve can still be manufactured from stainless steel. The actual structure of the valve varies a little from one manufacturer to another but all are modifications of the basic sample valve shown in figure 13. The valve usually consists of five parts. Firstly there is the control knob or handle that allows the valve selector to be rotated and thus determines the load and sample positions. Secondly, a connecting device that communicates the rotary movement to the rotor. Thirdly the valve body that contains the different ports necessary to provide connections to the mobile phase supply, the column, the sample loop if one is available, the sample injection port and finally a port to waste. Then there is the rotor that actually selects the mode of operation of the valve and contains slots that can connect the alternate ports in the valve body to provide loading and sampling functions. Finally there is a pre-load assembly that furnishes an adequate pressure between the faces of the rotor and the valve body to ensure a leak tight seal.

The Internal Loop Valve

As already stated, the valve system can take two basic forms, the internal loop sampling valve and the external loop sampling valve. A diagram of the internal loop valve which utilizes only four ports is shown in figure 14.

Figure 14 The Internal Loop Valve

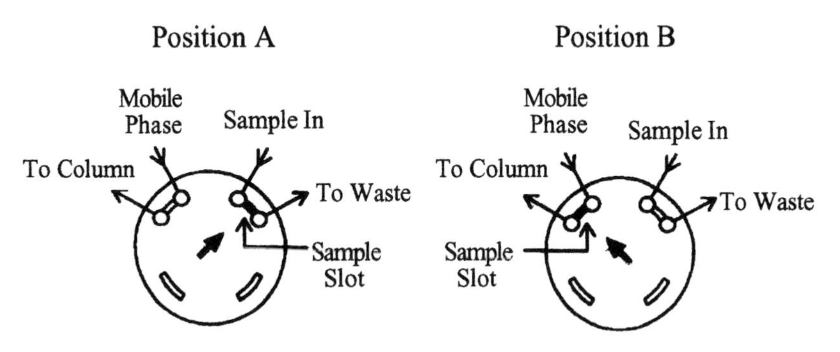

Courtesy of Valco Instruments Inc.

The sample volume of the internal loop valve is contained in the connecting slot of the valve rotor and thus is used only for relatively small sample volumes. Internal sample loop valves usually deliver samples ranging from 0.1 μl to about 0.5 μl. The operation of the valve is depicted in figure 14 and the diagram on the left-hand side shows the sample being loaded into the valve. The sample is occupying the rotor slot on the upper right and has been filled by passing the sample from an appropriate syringe through the rotor slot to waste. During this period the mobile phase supply has been passing through the valve and direct to the column. The valve is then rotated and the sample position is depicted on the right-hand side of figure 14. The valve slot containing the sample has now been imposed between the mobile phase supply and the column and, consequently, the sample is swept onto the column with the flow of solvent. A few seconds is allowed to pass to allow the sample to be completely transferred to the column and the valve can be returned to the load position. This type of valve is used for placing samples on columns where, due to the column design, the peak dispersion is constrained to a few microliters and thus the dispersion resulting from a significant sample volume would degrade the column performance. A typical example would be in the use of short columns, perhaps 4.6 mm in diameter and 3 cm long packed with support particles only 3 μ in diameter. Alternatively, they might also be used when employing relatively short columns (10 to 50 cm long) only about one millimeter or so in diameter.

The External Loop Valve

Where peak dispersion has not been constrained to very small volumes the external sample loop injector can be used and the external loop sample system, which employs six ports, is depicted in figure 15. In the external loop sample valve, three slots are cut in the rotor so that any adjacent pair of ports can be connected. In the loading position shown on the left, the mobile phase supply is connected by a rotor slot to port (4) and the column to port (5) thus allowing mobile phase to flow directly through the column. In this position the sample loop is connected to ports (3) and (6). Sample flows from a syringe into port (1) through the rotor slot to the sample loop at port (6). At the same time the third slot in the rotor connects the exit of the sample loop to waste at port (2).

Figure 15 The External Loop Valve

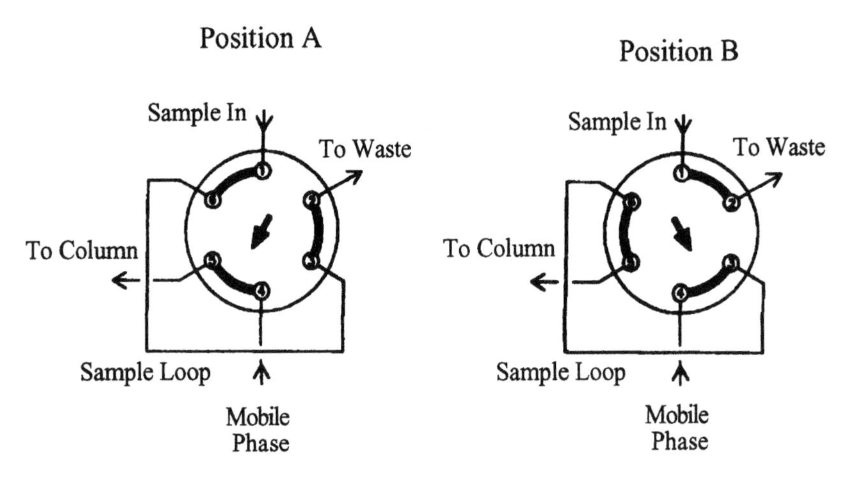

Courtesy of Valco Instruments Inc.

The sampling position is shown by the diagram on the right. On rotating the valve to place the sample on the column the sample loop is interposed between the column and the mobile phase supply by connecting port (3) and (4) and ports (5) and (6). Thus, the sample is swept onto the column. In the sampling position, the third rotor slot connects the syringe port to the waste port. After the sample has been placed on the column, the rotor can be returned to the loading position, the system washed with solvent and the sample loop loaded in readiness for the next injection. As already stated, the sample loop can take a wide range of volumes, those used for analysis usually falling within the range of 1 to 20 μl and those for preparative separations up to a 100 ml or more.

Column Switching

The technique of column switching can increase the versatility of the liquid chromatograph significantly. Examples of the use of column switching will be given in the chapter on applications but, as the technique employs valves similar to those used for sample injection the procedure will be discussed here. An example of a six port valve arranged for column switching is given in figure 16. The arrangement shown in figure 16 utilizes the same valve as that used for the external loop sampling system. It is seen that column (1) is connected between ports (5) and (6) and column (2) is connected between ports (2) and (3). Mobile phase from a sample valve, or more usually from another column, enters port (1) and the detector is connected to port (4). In the initial position of the rotor shown in the diagram on the left hand side, the rotor slots connect ports (1) and (6), (2) and (3) and (4) and (5).

Figure 16

Valve Arrangement for Column Switching

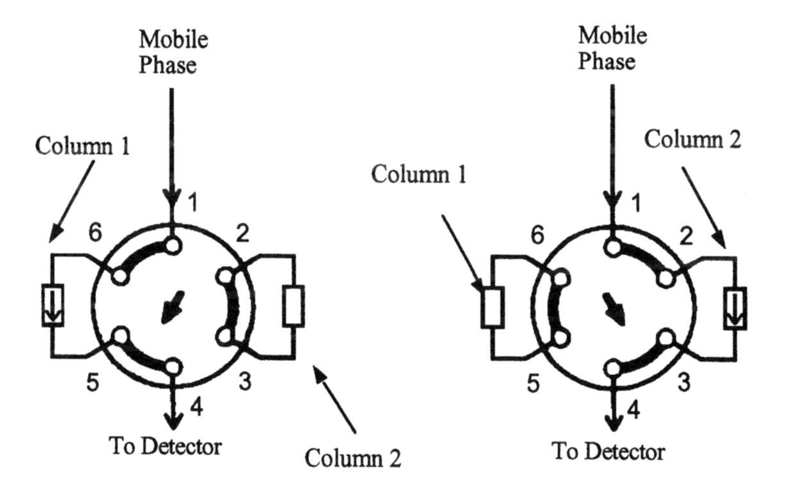

Courtesy of Valco Instruments Inc.

This results in mobile phase passing from port (1) to port (6), through column (1) to port (5), from port (5) to port (4) and out to the detector. Thus, the separation is taking place in column (1). The ports connected to column (2) are themselves connected by the third slot and thus isolated.

When the valve is rotated, the situation is depicted on the right hand side of figure (7); port (1) is connected to port (2), port (3) connected to port (4) and port (5) connected to port (6). This results in the mobile phase from either a sample valve or another column entering port (1) passing to port (2) through column (2) to port (3), then to port (4) and then to the detector. The ports (5) and (6) are connected, this time isolating column (1). This arrangement allows either one of two columns to be selected for an analysis or part of the eluent from another column pass to column (1) for separation and the rest passed to column (2). This system, although increasing the complexity of the column system renders the chromatographic process far more versatile. The number of LC applications that require such a complex chromatographic arrangement is relatively small; however, when required, column switching can provide a simple solution to difficult separation problems.

The Column and Column Oven

The use of temperature as a variable to control retention and selectivity has not been extensively employed in liquid chromatography. This is because, for most solutes, the change in free energy of a given solute in one phase with temperature is similar to that in the other phase. As a result there is little *net* change in retention and consequently little or no change in selectivity. Although, for some solutes, the *absolute* extent of solute retention can be affected by adjusting the temperature solute retention is more easily controlled by a change in mobile phase composition or a change in column length.

The effect of temperature on *column efficiency*, however, is frequently exploited, particularly in size exclusion chromatography (SEC). As has already been discussed, high efficiencies are essential in SEC due to the limited peak capacity of the column and consequently, the very small separation ratios. However the effect of temperature on column efficiency is not well understood by many analysts and consequently, will be discussed in some detail. It was shown on page

104, Chapter 4, from the rate theory that the variance per unit length of a column, or the HETP, was described by the following equation (1).

$$H = 2\lambda d_p + \frac{2\gamma D_m}{u} + \frac{f_1(k')d_p^2}{D_m}u + \frac{f_2(k')d_f^2}{D_S}u \quad (1)$$

where (λ) and (γ) are constants,
(D_m) and (D_S) are the diffusivities of the solute in the mobile and stationary phases respectively,
(d_p) is the particle diameter of the packing,
(d_f) is the effective thickness of the film of stationary phase,
(k') is the capacity ratio of the solute,
and (u) is the linear velocity of the mobile phase.

Now assuming that (k') is a constant (which, for a given solute, will be true in SEC) and a given column is considered, then the particle diameter and the film thickness are also constants. Furthermore, as the mobile phase and stationary phases are the same liquids (the stationary phase consists of the mobile phase held in the pores), then $D_m = D_S$.

Thus, equation (1) can be put in the form,

$$H = A' + \frac{B' D_m}{u} + \frac{(C_1 + C_2)}{D_m}u \quad (2)$$

Now as an increase in temperature will increase the value of (D_m) its effect on (H) and consequently the column efficiency is clear. If the mobile phase velocity is above the optimum then the function $\frac{(C_1 + C_2)}{D_m}u$ will dominate and an increase in temperature will increase (D_m), decrease $\frac{(C_1 + C_2)}{D_m}u$ and consequently (H) and thus *increase* the column efficiency. This is the most common situation in LC.

Conversely, if the mobile phase velocity is below the optimum velocity then the function $\frac{B' D_m}{u}$ will dominate and an increase in temperature will increase (D_m), increase $\frac{B' D_m}{u}$ and consequently (H) and thus *decrease* the column efficiency. LC columns are rarely operated below the optimum velocity and thus this situation is the least likely scenario.

It is interesting to ascertain the effect of temperature at the *optimum velocity* where the value of (H) is a minimum and the column efficiency a maximum.

Differentiating equation (2) with respect to (u) and equating to zero to find the optimum velocity,

$$\frac{\partial H}{\partial u} = -\frac{B' D_m}{u^2} + \frac{(C_1 + C_2)}{D_m}$$

Solving for the optimum velocity (u),

$$u_{opt} = D_m \sqrt{\frac{B'}{(C_1 + C_2)}} \qquad (3)$$

Substituting for u_{opt} from equation (3) in equation (1) a function for the minimum value of (H) is obtained.

$$H = A' + \frac{B' D_m}{D_m \sqrt{\frac{B'}{(C_1 + C_2)}}} + \frac{(C_1 + C_2)}{D_m} D_m \sqrt{\frac{B'}{(C_1 + C_2)}}$$

Simplifying,

$$H_{min} = A' + \sqrt{B'(C_1 + C_2)} \qquad (4)$$

It is seen that when operating at the optimum velocity that provides the minimum value of (H) and thus, the maximum efficiency, solute diffusivity has no effect on solute dispersion and consequently, the column efficiency is *independent of temperature*.

However, from equation (3) it is seen that the optimum velocity is directly related to the solute diffusivity and thus, to the temperature. It follows that increasing the temperature increases the optimum velocity and thus provides the same efficiency but a shorter analysis time.

Figure 17

An LC Column Oven Containing Column and Pre-Column

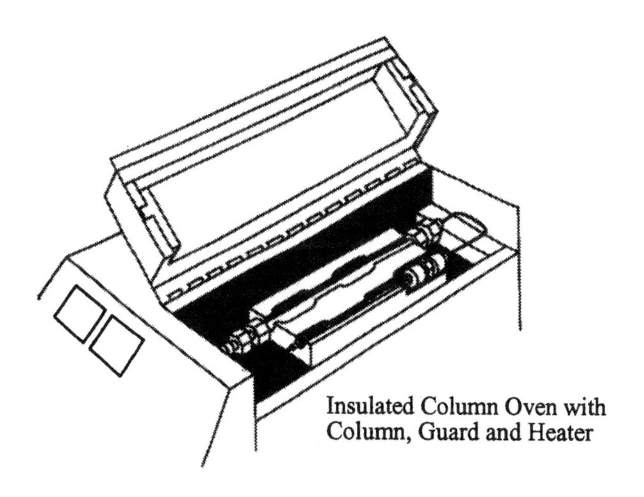

Courtesy of The Perkin Elmer Corporation

It follows that, at least for SEC, column temperature control can be important. An example of a commercially available column oven is shown in figure 17. The available temperature range varies a little from instrument to instrument but the model shown above has an operational range from 10°C to 99°C. One of the problems associated with the temperature control of ovens is the high thermal capacity of the column and that of the mobile phase. In general an air-oven will not be satisfactory due to the small heat capacity of gases generally. The use of liquid thermostatting media however, brings its own problems, including evaporation, spillage and, furthermore, column replacement and fitting tends to become a little messy. Perkin Elmer avoided this problem by using high thermal conductivity tubing to bring the mobile phase to thermal equilibrium with the oven temperature prior to entering the column. High thermal conductivity tubing is the same as low dispersion tubing (2) since the conduction of heat across a tube is mathematically equivalent to the treatment of solute transfer across a tube. The problem of heat or solute transfer across an open tube results from the parabolic velocity profile of the fluid passing through the tube and the relatively low magnitudes of solute diffusivity and the thermal conductivity of liquids.

Figure 18

Heat Transfer Tube

The diffusivity of a solute is that mass of solute transferred across unit area per unit time under a unit concentration gradient. Similarly the thermal conductivity of a liquid is the heat transferred across unit area of the liquid per unit time by a unit temperature gradient. To improve both the heat and solute transfer, it is necessary to break up the parabolic velocity profile in the tube and introduce radial mixing to enhance diffusion and thermal conductivity. The technique developed by Katz and Scott (2) employed a serpentine tube, the form of which is depicted in figure 18. For optimum transfer the ratio of the serpentine amplitude and the tube diameter should be less than 4. It functions by the introduction of radial flow as the fluid rapidly changes direction round the serpentine bends as it passes through the tube. The tube is clamped flat to the surface of the oven wall and the heat is transferred from the oven wall to the tube wall and then rapidly to the liquid passing through it. Without this, or some other type of efficient thermal exchange system, air ovens are not to be recommended for LC column temperature control.

Detectors

Excluding the column, the detector is the most important component of the liquid chromatograph. The accuracy and precision of the LC analysis depends crucially on the quality of the detector. As a result of its importance, the next chapter will be committed entirely to the discussion of detectors and so only certain aspects of detection will be considered here.

It is extremely important that the connection between the column and the detector be as short as possible and that the diameter of the conduit should be as small as is practical. Dispersion in open tubes can be extensive (3). The solute bands will spread in their passage between the column and detector and the resolution, so painstakingly achieved in the column itself, will be lost in transit through the connecting tube. Ideally, the column should be connected directly to the sensing cell of the detector but this, in most cases, is not mechanically possible. Nevertheless, the connection between column and detector in many commercial instruments may be 30 cm or more, which certainly shows poor instrument design and practice. Unfortunately, it is sometimes extremely difficult to determine the length of the connecting tubing due to the lay-out of the instrument. In fact, some manufacturers use a length of stainless steel tubing between the column and the detector as a heat exchanger to ensure temperature equilibrium in the detector cell. To determine the column detector dispersion, the sample valve should be connected directly to the detector, replacing the column with a short length (1 cm) of tubing with an i.d. of about 0.005 in. Using a 0.2 or 0.5 μl sample volume, a sample of a single solute (1% solution in the mobile phase) should be injected through the system at a flow rate of about 0.5 ml/min and the resulting peak concentration profile monitored. If the detector-column system is well designed the resulting peak should be symmetrical and the standard deviation of the peak (half the peak width) should be less than 3 μl (ca 0.4 sec at a flow rate of 1 ml/min). One of the most important specifications of a liquid chromatograph is the extra column dispersion (4) and this should be an important criteria when considering the purchase of an instrument. The importance of extra column dispersion can be seen by the following example.

A column 3 cm long, 4.6 mm in diameter packed with particles 3 μm in diameter will give about 6,000 theoretical plates at the optimum velocity. This efficiency is typical for a commercially available column.

Now the column volume (V_C) will be

$$V_C = \pi r^2 L = 3.14 \times 0.23^2 \times 3 = 0.499 \text{ ml}$$

where (r) is the column radius,
and (L) is the column length.

Now the dead volume (V_O) will be approximately 60% of the column volume and will be given by

$$V_o = 0.6V_c = 0.6 \times 0.499 = 0.299\text{ml}$$

From the Plate Theory already discussed, the standard deviation of the *dead volume* peak (σ_O) will be

$$\sigma_o = \frac{V_o}{\sqrt{n}} = \frac{0.299}{\sqrt{6000}} = 0.00386 \text{ ml} = 3.86 \ \mu\text{l}$$

It is clearly seen that the connecting tube dispersion of 3 μl will still be equivalent to 75% of the column dispersion. Ideally, the extra column dispersion, in terms of standard deviation, should only be about 1 μl but this is extremely difficult to achieve in practice. Some manufacturers have longer lengths of connecting tube but utilize the low dispersion tubing (2) already discussed, which can be a satisfactory alternative. In any event, the importance of knowing the magnitude of the extra column dispersion is abundantly clear. Unfortunately, in most cases, data for the extra-column dispersion of an instrument will not be available from the manufacturer and will need to be measured by the analyst, preferably before purchase.

The dimensions of the exit tube from the detector are not critical for analytical separations but they can be for preparative chromatography if fractions are to be collected for subsequent tests or examination. The dispersion that occurs in the detector exit tube is more difficult to measure. Another sample valve can be connected to the detector exit and the mobile phase passed backwards through the detecting system. The same experiment is performed, the same measurements made and the same calculations carried out. The dispersion that occurs in the exit tube is normally considerably greater than that between the column and the detector. However, providing the dispersion is known, the preparative separation can be adjusted to accommodate the exit tube dispersion and allow an accurate collection of each solute band.

Repetitive routine analysis of a specific sample (e.g., for Quality Control) will usually require a dedicated instrument. Therefore, the chromatograph and, in particular, the detector will be chosen for that specific analysis. Consequently, only one detector will be necessary and the purchase of an armory of detectors on the basis that they might be needed in the nebulous future is not advised. An alternative detector can always be obtained if and when the demand arises. The same argument applies to multi-solvent reservoirs and multi-solvent gradient programmers and other accessories that are not immediately required for the specific analysis in mind.

In contrast, an instrument required for method development must be as versatile as possible and all the alternative detectors and accessories should be available to reduce the time spent on the development procedure. Limited equipment can easily delay such research and, consequently, be very costly.

Data Acquisition and Processing

The analyst does not need to know the details of the electronic circuitry associated with the data acquisition system of a liquid chromatograph. In order to use the instrument effectively, however, a knowledge of the mathematical procedures that take place at each stage of the data processing will help identify the limits of accuracy and precision that can be expected from an analysis. The system employed differs slightly between different manufacturers, but that outlined below is generally representative of the procedure used. A block diagram showing the individual steps employed in acquiring and processing chromatographic data is given in figure 19.

Figure 19

A Typical Data Acquisition and Processing System

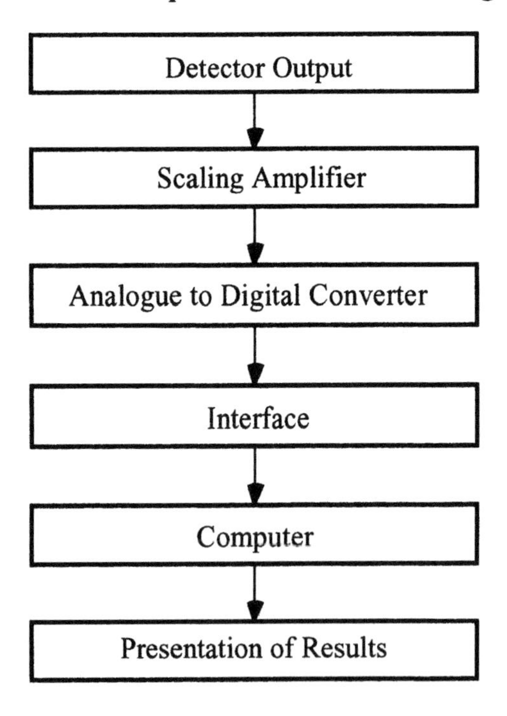

The output from the detector, which is usually in millivolts, is passed to a scaling amplifier that converts the voltage range of the detector to the range that is acceptable to the analog to digital (A/D) converter for example 0-1 volt. The A/D converter changes the voltage output from the scaling amplifier to a binary number which is temporarily stored in a register; this process is continuously repeated at a defined rate, called the 'sampling rate'. Each binary number, stored in the register is regularly read by the computer at the same frequency and stored. On completion of the chromatogram, the computer accesses all the data from store, calculates the retention report, compares peak heights or peak areas to provide the quantitative analysis according to the processing program that is used and finally prints out the results in tabulated form.

A simplified form of the acquisition procedure is shown diagramatically in figure 20. The output from most detectors range from zero to ten millivolts and the input range of many A/D converters is usually from zero to one volt. Thus, the instantaneous measurement of 0.2 mv from the detector must be scaled up by a factor of one hundred to 0.2 volts, which is carried out by the scaling amplifier. Now the A/D converter changes the analog voltage to a digital number, the magnitude of which is determined by the number of "bits" that the computer employs in its calculations. If for example eight bits are used the largest decimal number will be 255. The digital data shown in figure 20 can be processed backward to demonstrate A/D procedure. It is seen that the third and fourth most significant "bits" (which are counted from the far left) and the two least significant "bits" (which are counted from the far right) are at the five volt level, which as shown in figure 20 is equivalent to 51 in decimal notation (32+16+2+1). It follows that the voltage that was converted must be $\frac{51}{255}$ x 1 volt = 0.2 volt. It should also be noted that due to the limitation of 8 "bits", the minimum discrimination that can be made between any two numbers is $\frac{1}{255}$ x 100 $\approx$ 0.4%.

Figure 20

Stages of Data Acquisition

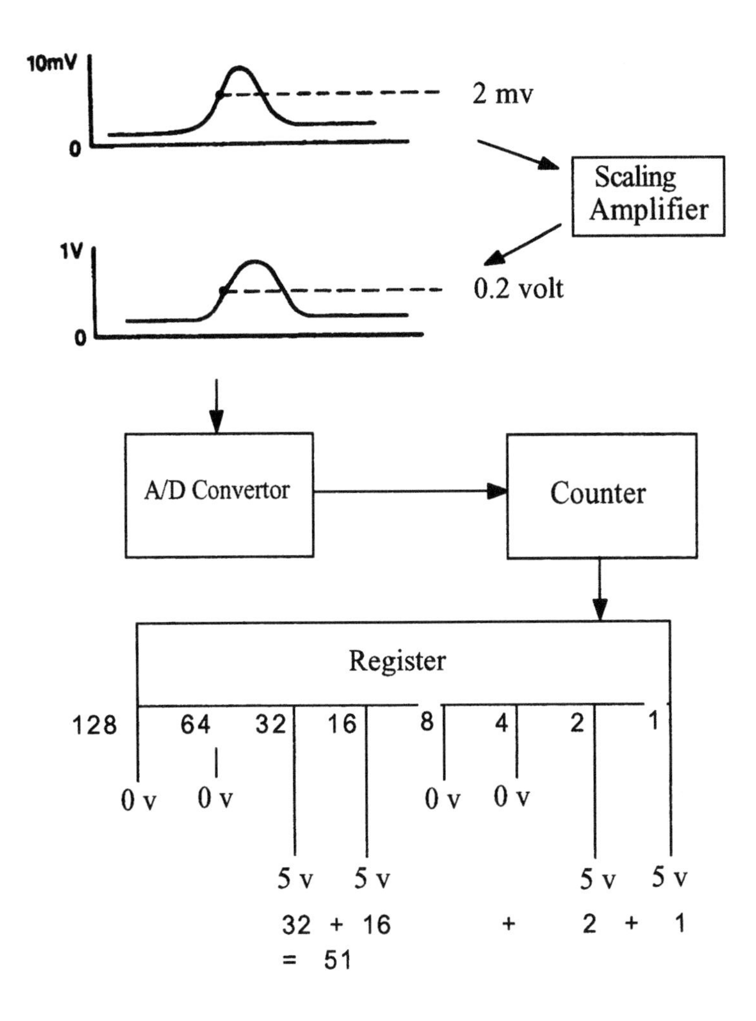

Thus the *number of bits* used for calculation *defines the minimum difference* between data values. Although 0.4 % appears a little high, other factors in the chromatographic procedure produce greater errors than the 0.4% resulting from the eight "bit" limitation. Many modern chromatography computer systems utilize more than eight "bits" in their data processing but, in practice, eight "bits" is often quite adequate due to the magnitude of other instrumental limitations on precision.

Modern data processing software includes routines that can be used to obtain accurate quantitative results from chromatograms where the components of the sample are incompletely resolved. The routines, in fact, deconvolute the individual peaks from the composite envelope and calculate the area of the individual de-convoluted peaks. Such algorithms can be used very effectively on peaks that are entrained in the tail of a major peak but are not so accurate for composite envelopes containing many unresolved peaks.

It should be emphasized that clever algorithms or subroutines are no substitute for good chromatography.

The chromatographic system should, wherever possible, be optimized to obtain complete resolution of the mixture and not place reliance on mathematical techniques to aid in the analysis.

References

1/ J.J. Van Deemter, F.J. Zuiderweg and A. Klinkenberg, *Chem. Eng. Sci.,* **5**(1956)271.

2/ E.D. Katz and R.P.W. Scott, *J.Chromatogr.*, **268**(1983)169.

3/"Small Bore Liquid Chromatography Columns", (Ed. R.P.W.Scott), John Wiley and Sons, New York-Chichester-Brisbane-Toronto-Singapore, (1984)23.

4/ R.P.W. Scott,"Liquid Chromatography Column Theory", John Wiley and Sons, New York-Chichester-Brisbane-Toronto-Singapore (1984)23.

6
Liquid Chromatography Detectors

The lack of a sensitive, linear detector was probably the greatest single impediment to the evolution of LC in the years past. The development of LC detectors has proved to be far more difficult than the development of GC detectors. This is because the presence of very low concentrations of solute vapor in a gas modify the physical properties of the gas to a far greater extent than the same concentration of solute will modify the physical properties of a liquid. Nevertheless, despite the sensitivity problems, many LC detectors have been developed over the past thirty years based on a variety of different sensing principles. However, only about twelve of them can be used effectively for LC analyses and, of those twelve, only four are in general use. The four prevailing detectors are the *UV detector* (fixed and variable wavelength), the *electrical conductivity detector*, the *fluorescence detector* and the *refractive index detector*; they are employed in over 95% of all LC analytical applications. In this chapter only these four detectors will be described. For those readers requiring more information on detectors, two books are recommended (1,2). Both books are dedicated to the subject of LC detectors.

Before describing the individual detectors, the subject of detector specifications must be discussed as the performance of the detector strongly influences the accuracy and precision that is realized in the LC analysis. There are a considerable number of detector specifications, many of which are crucial to the design engineer, but only seven are of prime importance to the analyst.

Detector Specifications

The seven major detector specifications are as follows:

Detector Linearity
Linear Dynamic Range
Detector Noise Level
Detector Sensitivity, or Minimum Detectable Concentration
Pressure Sensitivity
Flow Sensitivity
Temperature Sensitivity

Each of the above specifications will now be considered individually and those interested in the broader range of detector specifications are referred to reference (1).

Detector Linearity and Response Index (α)

True detector linearity is, in fact, a theoretical concept, and despite the claims by many manufacturers, LC detectors can only *tend* to exhibit this ideal response. As the linearity of the detector will determine the accuracy of the analysis, it is important to have some method for measuring detector linearity that can describe it in numerical terms. A method for linearity measurement was proposed by Scott and Fowlis (3), who assumed that for a nearly linear detector the response of the detector could be expressed by the following equation.

$$v = Rc^{\alpha} \qquad (1)$$

where (v) is the output from the detector,
(R) is a constant,
and (α) is the *response index* .

It is seen that, for a truly linear detector, the response index (α) will be unity and the experimentally determined value of (α) will be an accurate measure of the proximity of the response to strict linearity. It is also clear that (α) could be used to correct for any non-linearity that might occur in the detector and thus improve the accuracy of an analysis.

Taking logarithms, equation (2) becomes

$$\mathrm{Log}(v) = \mathrm{Log}(R) + \alpha \mathrm{Log}(c) \qquad (2)$$

It is seen that the value of (α) can be obtained from the slope of the curve relating the log (detector output) to the log (solute concentration) and an example of such a curve is shown in figure 1.

Figure 1

Graph of Log (Detector Output) / Log (Solute Concentration)

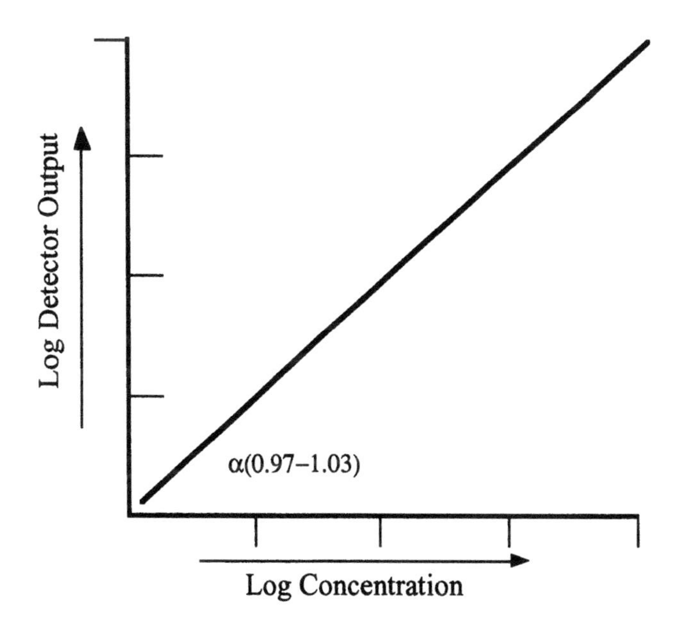

The curve can be obtained in practice by injecting samples having different concentrations of solute on to the column and measuring the area or height of the peaks for each injection. Each concentration should be run in duplicate and the average value used to construct the graph. If the results are being presented on a potentiometric recorder, then the peak *height* can be measured and corrected for the detector sensitivity settings as they are changed. If a computer data acquisition and processing system is used, then the peak *area* can be obtained directly from the print out and be used as an alternative to the peak height. If true linearity is to be assumed, then in order to obtain accurate quantitative results in LC analyses, the response index should lie between 0.98 and 1.02 as shown in figure 1. If the response index is not within this range then it will be necessary to correct for the non linearity employing the measured response index.

The curves relating detector output to solute concentration for detectors having different response indexes are shown in figure 2.

Figure 2

Graph of Detector Response against Solute Concentration

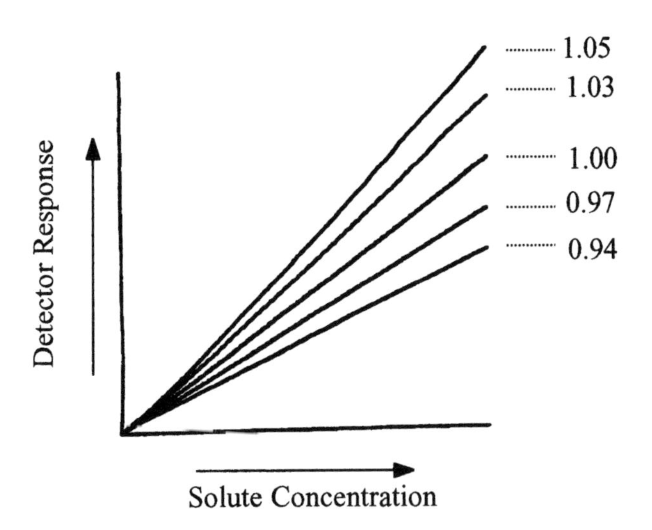

It is seen that, individually, all the curves appear linear and any one of the five curves might be considered to give accurate quantitative results. However, the actual results that would be obtained from the analysis of a binary mixture containing 10% of one component and 90% of the other, employing detectors with each of the five response factors, is shown in Table 1.

Table 1

Analysis of a Binary Mixture Employing Detectors with Different Response Indexes

Solute	α=0.94	α=0.97	α=1.00	α=1.03	α=1.05
1	11.25%	10.60%	10.00%	9.42%	9.05%
2	88.75%	89.40%	90.00%	90.58%	90.95%

It is seen that errors in the smaller component can be as great as 12.5% (1.25% absolute) when the response index is 0.94. Yet on examining the curve for a response index of 0.94 in figure 2 the non-linearity is scarcely apparent. When the response index is 1.05 the error is 9.5% (0.95% absolute) and again the poor linearity is not obvious in figure 2. As already stated, to obtain accurate results without employing a correction factor, *the response index should lie between 0.98 and 1.02.* Most LC detectors can be designed to meet this linearity criteria.

Linear Dynamic Range

As the linearity of a detector usually deteriorates at high solute concentration the *linear dynamic range* is not the same as the *dynamic range.* The *linear dynamic range* of a detector is that range of solute concentration over which the numerical value of the *repose index* also falls within a defined range. For example, the linear dynamic range of a detector might be specified as

$$D_L = 3 \times 10^{-8} \text{ to } 2 \times 10^{-5} \text{ g/ml } (0.98 < \alpha < 1.02)$$

The dynamic range of LC detectors is usually considerably less than their GC counterparts which evinces more care in determining sample size in quantitative analysis. A GC detector may have a linear response over a concentration range of five or six orders of magnitude, for example, the flame ionization detector, whereas an LC detector is more likely to have a dynamic range of only three orders of magnitude and some detectors considerably less.

Detector Noise Level

There are three different types of detector noise, short term noise, long term noise and drift. These sources of noise combine together to give the composite noise of the detector. The different types of noise are depicted in figure 3.

Figure 3

Different Types of Detector Noise

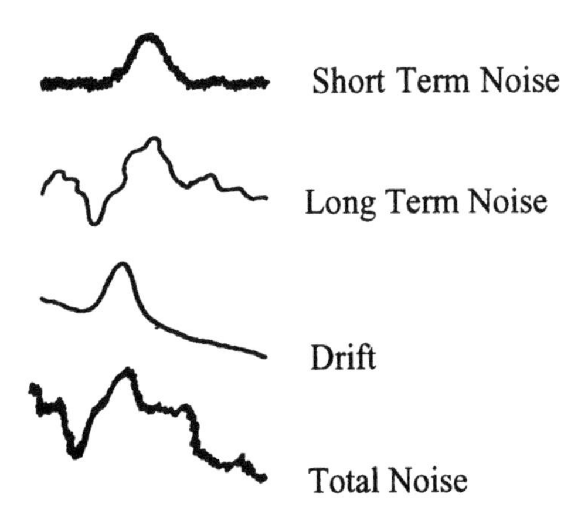

Short Term Noise consists of base line perturbations that have a frequency that is significantly higher than the eluted peak. Short term detector noise is not often a serious problem in liquid chromatography as it can be easily removed by an appropriate noise filter without affecting the profiles of the peaks. Its source is usually electronic, originating from either the detector sensor system or the amplifier.

Long Term Noise consists of base line perturbations that have a frequency that is similar to that of the eluted peak. This type of detector noise is the most serious as it is indiscernable from the small peaks and cannot be removed by electronic filtering without affecting the normal peak profiles. It is seen in figure 3 that the peak profile can easily be discerned above the high frequency noise but is lost in the long term noise. The source of long term noise is usually due to changes in either temperature, pressure or flow rate in the sensing cell. Long term noise is largely controlled by detector cell design and ultimately limits the detector *sensitivity* or the *minimum detectable concentration.*

Drift consists of base line perturbations that have a frequency that is significantly larger than that of the eluted peak. Drift is almost always due to either changes in ambient temperature, changes in solvent composition, or changes in flow rate. All these factors are easily constrained by careful control of the operating parameters of the chromatograph.

All three sources of noise combine to form the type of trace shown at the bottom of figure 3. In general, the sensitivity of the detector should never be set above the level where the combined noise exceeds 2% of the FSD (full scale deflection) of the recorder (if one is used) or appears as more than 2% FSD of the computer simulation of the chromatogram.

Measurement of Detector Noise

The detector noise is defined as the maximum amplitude of the combined short and long term noise, measured in millivolts, over a period of fifteen minutes. If a column 4.5 mm i.d. is employed, a flow rate of 1 ml/min is appropriate. The flow rate should be adjusted appropriately for columns of different diameters. The value for the detector noise should be obtained by constructing parallel lines embracing the maximum excursions of the recorder trace over the defined time period as shown in figure 4. The distance between the parallel lines measured in millivolts is taken as the noise level.

Figure 4

Method for Measuring of Detector Noise

Detector Sensitivity or the Minimum Detectable Concentration

Detector Sensitivity or the Minimum Detectable Concentration has been defined as the minimum concentration of an eluted solute that can be differentiated unambiguously from the noise. The ratio of the signal to the noise for a peak that is considered decisively identifiable has been arbitrarily chosen to be two. This ratio originated from electronic theory and has been transposed to LC. Nevertheless, the ratio is realistic and any peak having a signal to noise ratio of less than two is seriously obscured by the noise. Thus, the minimum detectable concentration is that concentration that provides a signal equivalent to twice the noise level. Unfortunately, the concentration that will provide a signal equivalent to twice the noise level will usually depend on the physical properties of the solute used for measurement. Consequently, the detector sensitivity, or minimum detectable concentration, must be quoted in conjunction with the solute that is used for measurement.

Pressure Sensitivity

The pressure sensitivity of a detector will be one of the factors that determines the long term noise and thus can be very important. It is usually measured as the change in detector output for unit change in sensor-cell pressure. Pressure sensitivity and flow sensitivity are to some extent interdependent, subject to the manner in which the detector functions. The UV detector, the fluorescence detector and the electrical conductivity detector can be designed to be relatively insensitive to pressure changes on the sensing cell, but the refractive index detector is very responsive to fluctuations in cell pressure. It follows that when selecting a refractive index detector the Pressure Sensitivity will be an important specification to consider.

Flow Sensitivity

The flow sensitivity of a detector will also be one of the factors that determines the long term noise and thus will influence the sensitivity or minimum detectable concentration of the detector.It is usually measured as the change in detector output for unit change in flow rate through the sensor cell. Again, the refractive index detector is the most sensitive to flow rate changes.

Temperature Sensitivity

Both the sensing device of the LC detector and the associated electronics can be temperature sensitive and cause the detector output to drift as the ambient temperature changes. Consequently, the detecting system should be designed to reduce this drift to a minimum. In practice the drift should be less than 1% of FSD at the maximum sensitivity for $1^{o}C$ change in ambient temperature.

The UV Detector

The UV detector is the most popular and useful LC detector that is available to the analyst at this time. This is particularly true if multi-wavelength technology is included in the genus of UV detectors. Although the UV detector has definite limitations, particularly with respect to the detection of non-polar solutes that do not possess a UV chromaphore, it has the best combination of sensitivity, versatility and reliability of all the detectors so far developed for general LC analyses.

The majority of compounds adsorb UV light in the range of 200-350Å including all substances having one or more double bonds (π electrons) and all substances that have unshared (non bonded) electrons (e.g., all olefins, all aromatics and all substances containing >CO, >CS, -N=O and $-N \equiv N-$ groups). The relationship between the intensity of UV light transmitted through the cell and solute concentration is given by Beer's Law,

$$I_T = I_o e^{-kLc}$$

where (I_O) is the intensity of the light entering the cell,
(I_T) is the intensity of the transmitted light,
(L) is the path length of the cell,
(c) is the concentration of the solute,
(k) is the molar extinction coefficient of the solute for the specific wavelength of the UV light.

or

$$\text{Ln } I_T = \ln I_o - kLc$$

Rearranging,

$$I_T = I_o 10^{-k'Lc}$$

where (k') is the molar extinction coefficient of the solute.

Differentiating,

$$\frac{\partial\left(\text{Ln}\frac{I_T}{I_o}\right)}{\partial c} = -kL \qquad (3)$$

Equation (3) displays the factors that control the detector sensitivity. It is seen that the sensitivity of the detector is controlled by the magnitude of the extinction coefficient for the solute being measured, which, in turn, will depend on the wavelength of the UV light that is used. It follows that the minimum detectable concentration can be changed by selecting a light source of different wavelength. The detector sensitivity is seen to depend also on the path length of the cell. Unfortunately, the cell length cannot be increased indefinitely to provide higher sensitivity as long cells will provide excessive peak dispersion with consequent loss of column resolution. It follows that the optimum detector cell design involves the determination of the cell length that will provide the maximum sensitivity and at the same time constrain detector dispersion to a minimum so that there is minimum loss in resolution.

The Fixed Wavelength Detector

There are two types of UV detector: the *fixed wavelength detector* and the *multi-wavelength detector*. A diagram of a fixed wavelength UV detector is shown in figure 5.

Figure 5 The Fixed Wavelength UV Detector

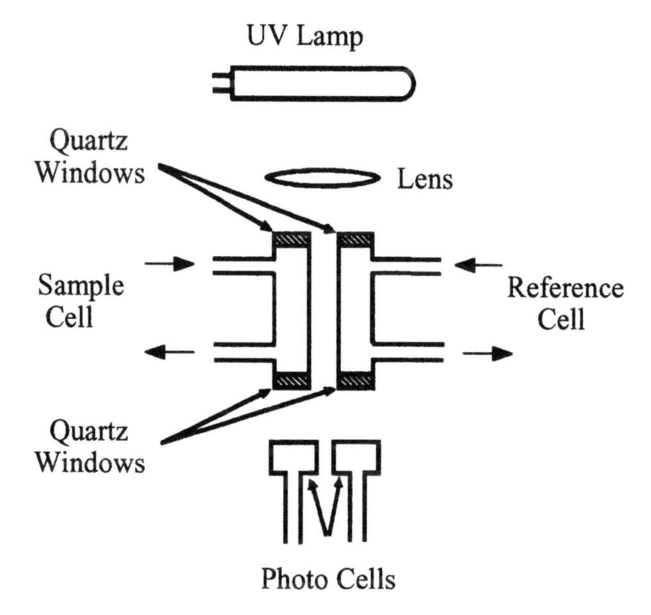

The detector consists of a small cylindrical cell (2.0 to 10.0μl in volume) through which flows the eluent from the column. UV light from an appropriate UV lamp passes through the cell and falls on a UV photo-electric cell. In a fixed wavelength detector the wavelength of the light depends on the type of lamp that is used. There are a number of lamps available that provide UV light having wavelengths ranging from about 210 nm to 280 nm. The lamps that are commercially available at the time of writing this book are as follows:

Lamp Type	**Emission Wavelengths**
Mercury Vapor Lamp	253.7 nm
Zinc Vapor Lamp	212.9 nm and 307.6 nm
Cadmium Vapor Lamp	228.8, 326.1, 340.3, and 346.6 nm

The mercury vapor lamp is by far the most popular, having an emission wavelength that gives the detector the maximum versatility for detecting a wide range of solute types. The detector usually contains a sample and reference cell and the output from the reference cell is compared to that from the sample cell. The difference is fed to a non-linear amplifier that converts the signal to one that is linearly related to concentration of solute in the sample cell. The fixed wavelength detector is the least expensive and, as all the light is emitted at a specific wavelength(s), it has a higher intrinsic sensitivity than the multi-wavelength UV detectors. However, the multi-wavelength detector can often compensate for this lack of sensitivity by operating at a wavelength where the absorption of the solute is a maximum and thus, provides the greatest response. Average specifications for commercially available fixed wavelength UV detectors are as follows:

Typical Specifications for a Fixed Wavelength UV Detector

Sensitivity (toluene)	5×10^{-8} g/ml
Linear Dynamic Range	5×10^{-8} to 5×10^{-4} g/ml
Response Index	0.98 - 1.02

By the use of very small sensing cells and electronic systems with very small time constants, the fixed wavelength detector can be designed to give a very fast response at high sensitivity and very low dispersion and for this reason it can be used for very fast analyses. An example of the rapid separation of a two-component mixture (4) is shown in figure 6.

The reason for separating benzene and benzyl acetate in 2.6 seconds remains (to say the least) obscure and figure 6 is obviously an example of "Chromatography Show Biz". Nevertheless, it does demonstrate that columns can be designed and detectors developed that can provide extremely fast analyses.

Figure 6

The High Speed Separation of a Two Component Mixture

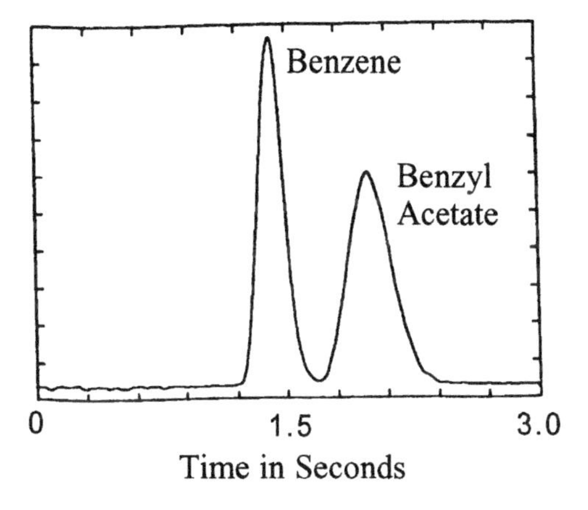

The Multi-Wavelength Detector

Multi-wavelength UV detectors utilize a single, or, more accurately in practice, a narrow range of wavelengths to detect the solute. Most UV detectors can also provide a UV spectra of the eluted sample if appropriately arranged. There are basically two types of multi-wavelength detectors: the *dispersion detector,* that monitors the eluent at one wavelength only, and the *diode array detector,* that simultaneously monitors the eluted solute over a range of wavelengths. The former passes the light from a broad emission light source through a monochromator, selects a specific wavelength and allows it to pass through the detecting cell. The second also uses a broad emission light source, but all the light is allowed to pass through the sensing cell and *subsequently* the light is dispersed by means of a holographic grating and the dispersed light allowed to fall on an array of diodes. The two alternative types of detectors will be discussed separately.

The Multi-Wavelength Dispersive Detector

A diagram of the Multi-Wavelength Dispersive Detector is shown in figure 7.

Figure 7

The Multi-Wave Length Dispersive Detector

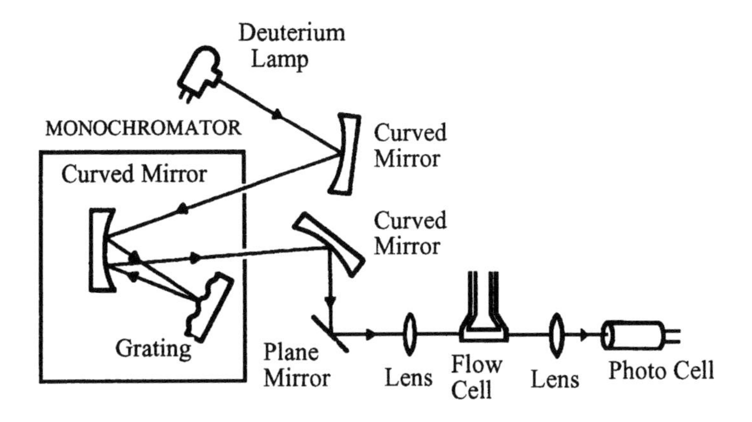

Light from a broad wavelength emission source, such as a deuterium or xenon discharge lamp, is collimated by two curved mirrors onto a holographic diffraction grating. The dispersed light is focused by means of a curved mirror on to a plane mirror and light of specific wavelength selected by appropriately positioning the angle of the plane mirror. Light of the selected wavelength is then focused by means of a lens through the flow cell and, consequently, through the column eluent. The exit beam from the cell is focused by another lens onto a photo cell which gives a response that is some function of the intensity of the transmitted light. The detector is usually fitted with a scanning facility that, by arresting the flow of mobile phase, allows the spectrum of the solute contained in the cell to be obtained. Due to the limited information given by most UV spectra and the similarity between many spectra of widely different types of compound, this procedure is not very reliable for the identification of most solutes. The technique is useful, however, for determining the homogeneity of a peak by obtaining spectra from a sample on both sides of the peak. The technique is to normalize both spectra, then either subtract one from the other and show that the difference is close to zero, or take the ratio and show it is constant throughout the peak.

A more common use of the multi-wavelength detector is to select a wavelength that is characteristically absorbed by a particular component or components of a mixture. This can be done to enhance either the sensitivity of the detector to those particular solutes, or to render the detector more specific and thus not respond significantly to substances of little interest in the mixture. The multi-wavelength dispersive detector is probably the most useful type of UV detector, providing adequate sensitivity, versatility and a linear response. It is however somewhat bulky (due to the need for a relatively large internal 'optical bench'), has *mechanically operated* wavelength selection and requires a stop/flow procedure to obtain spectra "on-the-fly". The diode array detector has the same advantages but none of the disadvantages athough, as one might expect, it is somewhat more expensive.

The Diode Array Detector

The diode array detector, although offering detection over a range of UV wavelength, functions in an entirely different way from the dispersive instrument. A diagram of a diode array detector is shown in figure 8. Light from the broad emission source such as a deuterium lamp is collimated by an achromatic lens system so that the total light passes through the detector cell onto a holographic grating. In this way the sample is subjected to light of all wavelengths generated by the lamp.

Figure 8

The Diode Array Detector

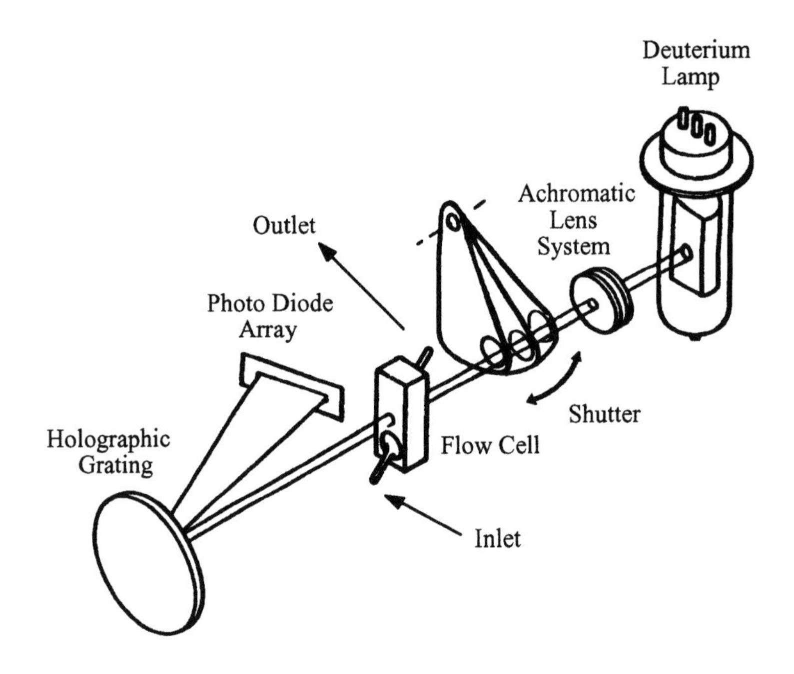

The dispersed light from the grating is allowed to fall on to a diode array. The array may contain many hundreds of diodes and the output from each diode is regularly sampled by a computer and stored on a hard disc. At the end of the run, the output from any diode can be selected and a chromatogram produced employing the UV wavelength that was falling on that particular diode. Most instruments will permit the monitoring of at least one diode in real time so that the chromatogram can be followed as the separation develops. This system is ideal, in that by noting the time of a particular peak, a spectrum of the solute can be obtained by recalling from memory the output of all the diodes at that particular time. This gives directly the spectrum of the solute i.e. a curve relating adsorption against wavelength.

The diode array detector can be used in a number of unique ways and an example of the use of a diode array detector to verify the purity of a given solute is shown in figure 9.

Figure 9

Dual Channel Plot from a Diode Array Detector Confirming Peak Purity

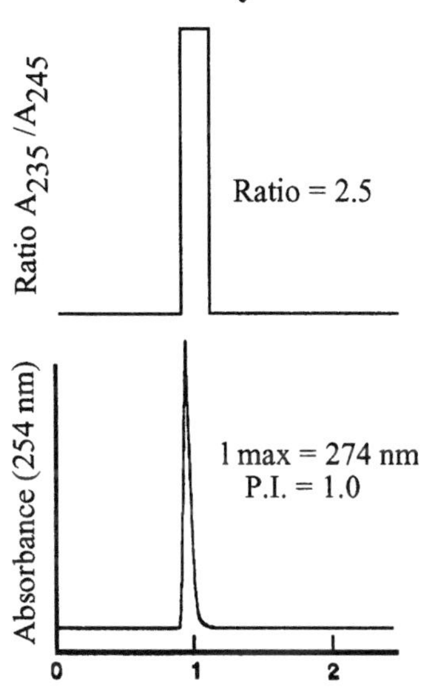

Courtesy of the Perkin Elmer Corporation

The chlorthalidone was isolated from a sample of tablets and separated by a reverse phase (C18) on a column 4.6 mm i.d., 3.3 cm long, using a solvent mixture consisting of 35% methanol, 65% aqueous acetic acid solution (water containing 1% of acetic acid). The flow rate was 2 ml/min and the chromatogram produced monitored at 274 nm is shown in the lower part of figure 9. As a diode array detector was employed, it was possible to ratio the output from the detector at different wavelengths and plot the ratio simultaneously with the chromatogram monitored at 274 nm. Now, if the peak was pure and homogeneous, the ratio of the adsorption at the two wavelengths (those selected being 225 and 245 nm) would remain constant throughout the elution of the entire peak. The upper diagram in figure (9) shows this ratio plotted on the same time scale and it is seen that a clean rectangular peak is observed which unambiguously confirms the purity of the peak for chlorthalidone. It should be pointed out that the wavelength chosen to provide the confirming ratio will depend on the UV adsorption characteristics of the substance concerned, relative to those of the most likely impurities to be present.

Another interesting example of the use of the diode array detector to confirm the integrity of an eluted peak is afforded by an application published by the Perkin Elmer Corporation showing the separation of a mixture of aromatic hydrocarbons. The separation they examined is shown in figure 10.

Figure 10

The Separation of a Series of Aromatic Hydrocarbons

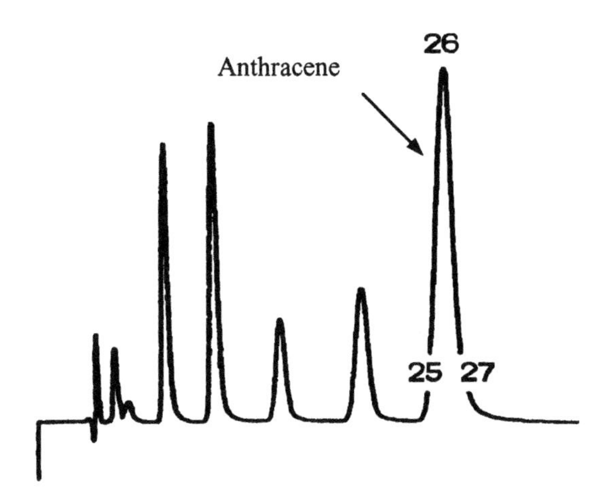

Courtesy of the Perkin Elmer Corporation

The separation was carried out on a column 3 cm long, 4.6 mm in diameter and packed with a C18 reversed phase on particles 3 μ in diameter. It is seen that the separation appears to be good and all the peaks represent individual solutes. However, by plotting the adsorption ratio, $\frac{250\,\text{nm}}{255\,\text{nm}}$, for the anthracene peak it was clear that the peak tail contained an impurity. The absorption ratio peaks are shown in figure 11.

Figure 11

Curves Relating the Adsorption Ratio, $\frac{250\,\text{nm}}{255\,\text{nm}}$, against Time

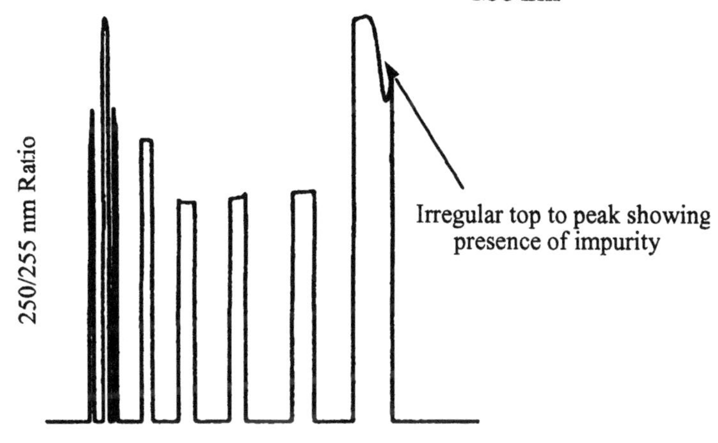

Spectra taken at the leading and trailing edge of the anthracene peak are shown superimposed in figure 12.

Figure 12

Superimposed Spectra Taken at the Leading and Trailing Edges of the Anthracene Peak

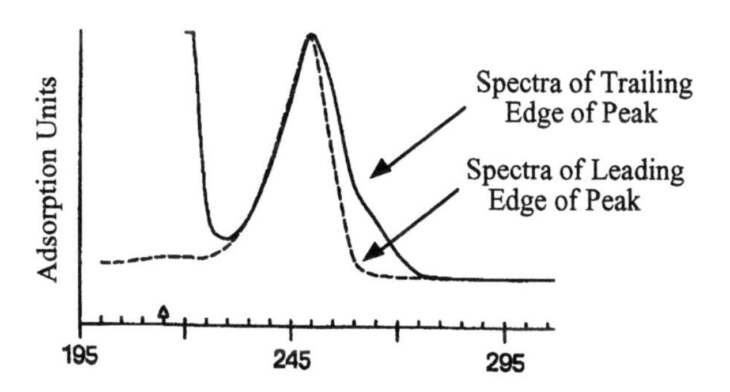

The ratio peaks in figure 11 clearly show the presence of an impurity by the sloping top of the anthracene peak and again this is confirmed unambiguously by the difference in the spectra for the leading and trailing edges of the peak. Further work identified the impurity as t-butyl benzene at a level of about 5%.

The versatility and advantages of the diode array detector are obvious but it is basically a research instrument or, from the point of view of the analyst, would be extremely useful in method development. Its use in routine analysis, however, might be considered vernacularly as "overkill". In any routine analysis, its versatility would be hardly used and its expense might be difficult to justify.

The performance of both types of multi-wavelength detectors are very similar and typical values for their more important specifications are as follows:

Typical Specifications for a Multi-Wavelength UV Detector

Sensitivity	1×10^{-7} g/ml
Linear Dynamic Range	5×10^{-7} to 5×10^{-4} g/ml
Response Index	0.97 - 1.03

The Electrical Conductivity Detector

The electrical conductivity detector is probably the second most commonly used in LC. By its nature, it can only detect those substances that ionize and, consequently, is used frequently in the analysis of inorganic acids, bases and salts. It has also found particular use in the detection of those ionic materials that are frequently required in environmental studies and in biotechnology applications. The detection system is the simplest of all the detectors and consists only of two electrodes situated in a suitable detector cell. An example of an electrical conductivity detector sensing cell is shown in figure 13.

Figure 13

An Electrical Conductivity Detector Sensing Cell

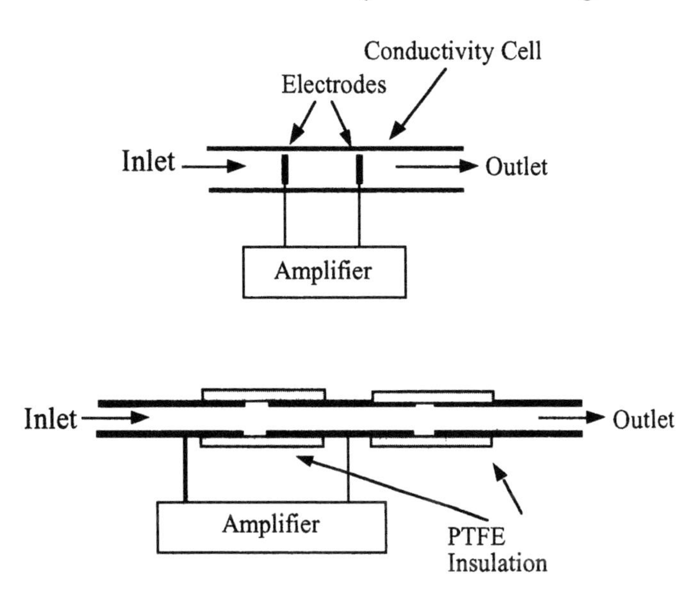

Basically, the detector consists of two electrodes situated in a suitable cell through which the column eluent flows. The theoretical configuration is depicted in the upper diagram. Electronically, the electrodes are arranged to constitute one arm of a Wheatstone Bridge. When ions enter the detector cell, the electrical resistance changes and the out-of-balance signal is fed to a suitable amplifier. The output from the amplifier is either digitized, and the binary number sent to a computer for storage, or the output is passed directly to a potentiometric recorder. The detector actually measures the electrical resistance between the electrodes which, by suitable non-linear amplification, can be made to provide an output that is linearly related to solute concentration. It should be pointed out that, to avoid electrode polarization, it is essential that an AC voltage is used across the electrodes to measure the cell impedance. The frequency of the AC potential across the electrodes is usually around 10 kHz.

A more practical system is shown in the lower part of the diagram The cell consists of short lengths of stainless steel tube, insulated from each other by a PTFE connecting sleeve. For convenience, in the design of the supporting electronic system, the first tube that is connected to the column is usually grounded (earthed). The resistance between the inlet tube and the center tube is continuously monitored, which, as is seen from the diagram, will be the resistance across the tiny gap between the tubes contained in the first PTFE sleeve. The volume of eluent in this tiny gap can be made extremely small and thus, providing tubes of small diameter and length are employed, the peak dispersion that takes place in an electrical conductivity cell can also be made exceedingly small. The resistance of the solution situated between the tubes is inversely proportional to the electric conductivity of the solution which, in turn, is related to the ion concentration in mobile phase. Some typical specifications for an electrical conductivity detector are as follows:

Typical Specifications for an Electrical Conductivity Detector

Sensitivity	5×10^{-9} g/ml
Linear Dynamic Range	5×10^{-9} to 1×10^{-6} g/ml
Response Index	0.97 - 1.03

An example of the use of the electrical conductivity detector is given by the separation and determination of a mixture of alkali and alkaline earth cations at levels of a few parts per million. The cations lithium, sodium, ammonium, potassium, magnesium and calcium were present in the original mixture at concentrations of 1, 4, 10, 10, 5 and 10 ppm respectively. The separation obtained is shown in figure 14. A proprietary ion exchange column, IonPacCS12, was used and the mobile phase consisted of a 20 nM methanesulphonic acid solution in water. A flow rate of 1 ml/min was employed and the sample volume was 25 µl. The separation is also an interesting example of the use of the *ion suppression* technique. The methanesulphonic acid solution, if passed through the detector, would have had a high electrical conductivity and thus, give a large signal on the detector which would swamp the signal from the ions being determined. Thus, after the mobile phase leaves the column and consequently, after the methane sulphonic acid has achieved its purpose and helped produce the desired separation, the agent must be removed.

Figure 14

Determination of Alkali and Alkaline Earth Cations

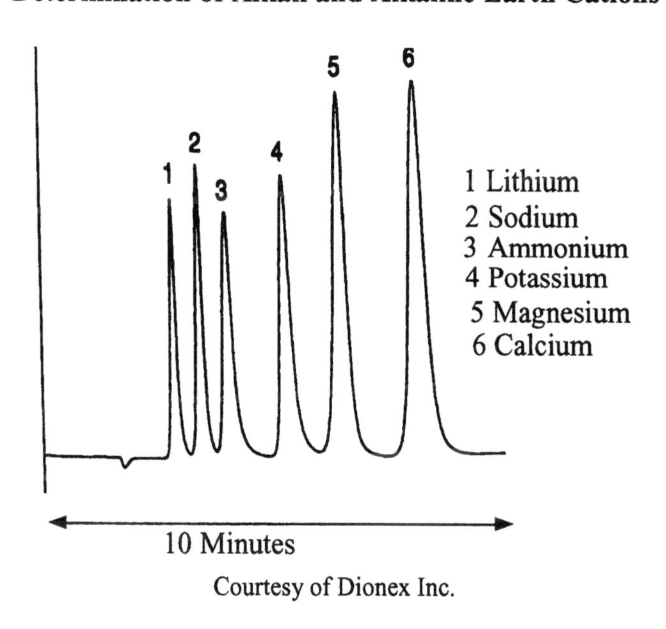

Courtesy of Dionex Inc.

This is necessary so that the mobile phase actually entering the detector contains little or no ions other than those being analyzed and so there is minimal background conductivity. The methane sulphonic acid can be removed by a number of procedures, but probably the simplest is to pass the mobile phase, after it has left the column, through a short reverse phase column. The reverse phase column, as already discussed, will remove any organic material by adsorption due to the strong dispersive forces that will occur between the hydrocarbon chains of the reverse phase and the methyl group of the methanesulphonic acid. Obviously the ion suppression column will eventually saturate and require regeneration. This can be easily achieved by desorbing the methane sulphonic acid with a strong dispersive solvent that is miscible with water such as acetonitrile. This technique of ion suppression is frequently used in ion exchange chromatography when using the electrical conductivity detector. However, depending on the nature of the separation, a specific type of ion suppression column or membrane will be required that will be appropriate for the phase system that is chosen. A wide variety of different types of ion suppression columns are available for this purpose. It should be pointed out that any supressor system introduced between the column and the detector that has a finite volume might cause band spreading. Consequently, the connecting tubes and suppression column must be very carefully designed to eliminate or reduce this dispersion to an absolute minimum.

The Fluorescence Detector

The fluorescence detector is probably the most sensitive of all LC detectors and as a result is often used for trace analysis. When molecules are excited by electromagnetic radiation and produce luminescence the effect is called photoluminescence. If the release of electromagnetic energy is immediate and stops upon the removal of the exciting radiation, the effect is called fluorescence and it is this phenomenon that is utilized in the fluorescence detector. Unfortunately, although the detector is very sensitive, its response is linear over a relatively limited concentration range. In fact, for accurate work, the detector cannot be assumed to be linear over a concentration range of much more than two orders of magnitude. Another disadvantage to the detector is that the majority of substances do not naturally fluoresce and thus fluorescent derivatives of the substances must be synthesized to render them detectable. There are a number of regents that have been developed specifically for this purpose but the derivatizing procedure will be considered in detail when methods of sample preparation are being discussed. A diagram of the fluorescence detector is shown in figure 15.

In the simplest form, light from a fixed wavelength UV lamp passes through a cell, through which the column eluent flows, and the fluorescent light that is emitted by a solute is sensed by a photoelectric cell positioned normal to the direction of exciting UV light.

Figure 15

The Fluorescence Detector

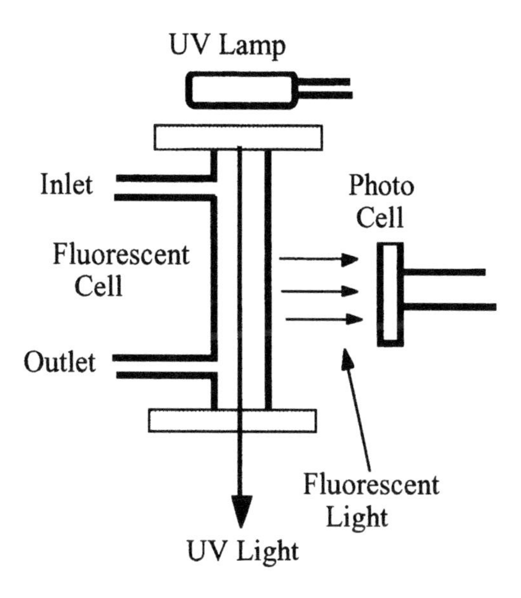

The photo cell senses light of all wavelengths that is generated by fluorescence but the wavelength of the excitation light can only be changed by use of an alternative lamp. This simple type of fluorescence detector was the first to be developed, is relatively inexpensive, and for certain compounds can be extremely sensitive. Typical specifications for a fluorescence detector are as follows:

Typical Specifications for a Fluorescence Detector

Sensitivity (Anthracene)	1×10^{-9} g/ml
Linear Dynamic Range	$1 \times 10^{-9}$to 5×10^{-6} g/ml
Response Index	0.96 - 1.04

The ultimate in fluorescence detection is a detector that uses a monochromator to select the excitation wavelength and a second monochromator to select the wavelength of the fluorescent light. This instrument is ideal, giving the maximum versatility and allowing the maximum sensitivity to be realized for any solute or solute type. An excellent example of an application of this type of complex system is afforded by the separation of the fifteen priority pollutants using programmed fluorescence detection. A chromatogram of a sample of this type is shown in figure 16.

Figure 16

Separation of a Series of Priority Pollutants with Programmed Fluorescence Detection

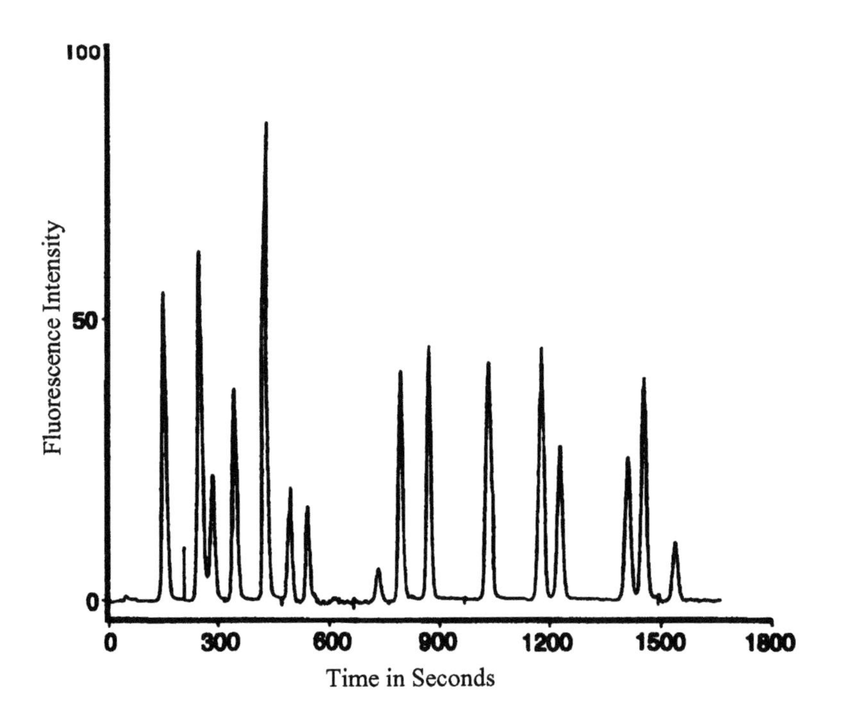

Courtesy of the Perkin Elmer Corporation

The separation was carried out on a column 25 cm long, 4.6 mm in diameter and packed with a C18 reversed phase. The mobile phase was programmed from a 93% acetonitrile 7% water to 99% acetonitrile 1% water over a period of 30 minutes. The gradient was linear and the flow rate was 1.3 ml/min. All the solutes are separated and the compounds, numbering from the left are as follows:

Fifteen Priority Pollutants

1 Naphthalene
2 Acenaphthene
3 Fluorene
4 Phenanthrene
5 Anthracene
6 Fluoranthene
7 Pyrene
8 Benz(a)anthracene
9 Chrysene
10 Benzo(b)fluoranthene
11 Benzo(k)fluoranthene
12 Benzo(a)pyrene
13 Dibenz (a,h)anthracene
14 Benzo(ghi)perylene
15 Indeno(123-cd)pyrene

The program used for the fluorescence detector was complex as, during the development of the separation, both the wavelength of the excitation light and the wavelength of the emission light that was monitored was changed. The program that was used is shown below.

Fluorescence Detector Program

Time (seconds)	Wavelength of Excitation Light	Wavelength of Emitted Light
0	280 nm	340 nm
220	290 nm	320 nm
340	250 nm	385 nm
510	260 nm	420 nm
720	265 nm	380 nm
1050	290 nm	430 nm
1620	300 nm	500 nm

It is seen that the analysis is an elaborate procedure carried out with a very complex and expensive instrument. Nevertheless, if the analysis is sufficiently important it may be necessary to resort to this solution. The system can also provide a fluorescence spectra, should it be required, by arresting the flow of mobile phase when the solute resides in the detecting cell and scanning the fluorescent light.

The Refractive Index Detector

The refractive index detector is the least sensitive of all the commercially available and generally useful detectors. It is very sensitive to changes in ambient temperature, pressure changes, flow-rate changes and can not be used for gradient elution. Nevertheless this detector is extremely useful for detecting those compounds that are nonionic, do not adsorb in the UV and do not fluoresce. There are a number of optical systems used in refractive index detectors (1), but the simplest and most common in use is the differential refractive index detector shown diagramatically in figure 17.

Figure 17

The Refractive Index Detector

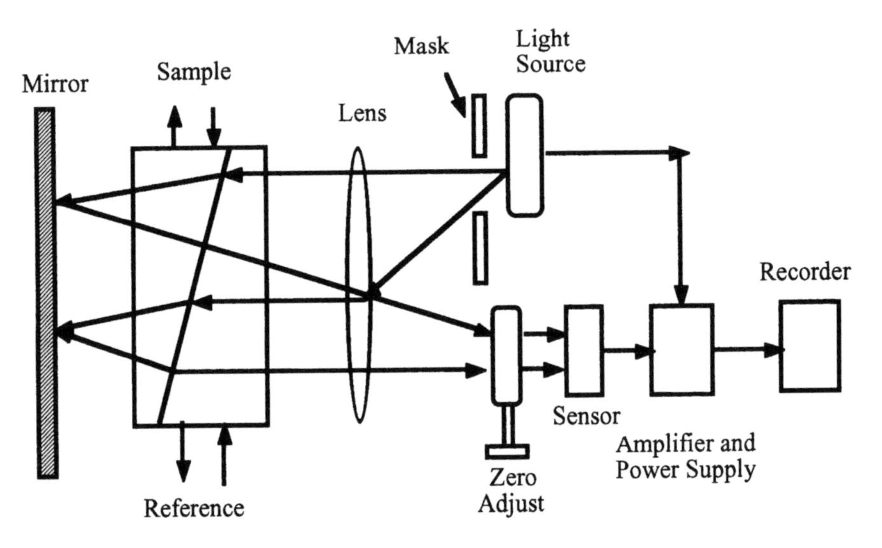

Courtesy of the Millipore Corporation

The differential refractometer monitors the deflection of a light beam caused by the difference in refractive index between the contents of the sample cell and those of the reference cell. A beam of light from an incandescent lamp passes through an optical mask that confines the beam to the region of the cell. The lens collimates the light beam which passes through both the sample and reference cells to a plane mirror. The mirror reflects the beam back through the sample and reference cells to a lens which focuses it onto a photo cell. The location of the beam, rather than its intensity, is determined by the angular deflection of the beam caused by the refractive index difference between the contents of the two cells. As the beam changes its position of focus on the photoelectric cell, the output changes and the resulting difference signal is electronically modified to provide a signal proportional to the concentration of solute in the sample cell.

The refractive index detector, in general, is a 'choice of last resort' and is used for those applications where, for one reason or another, all other detectors are inappropriate or impractical. However, the detector has one particular area of application for which it is unique and that is in the separation and analysis of polymers. In general, for those polymers that contain more than six monomer units, the refractive index is directly proportional to the concentration of the polymer and is practically independent of the molecular weight. Thus, a quantitative analysis of a polymer mixture can be obtained by the simple normalization of the peak areas in the chromatogram, there being no need for the use of individual response factors. Some typical specifications for the refractive index detector are as follows:

Typical Specifications for a Refractive Index Detector

Sensitivity (benzene)	1×10^{-6} g/ml
Linear Dynamic Range	1×10^{-6} to 1×10^{-4} g/ml
Response Index	0.97 - 1.03

A typical application of the RI detector is in carbohydrate analysis. Carbohydrates do not adsorb in the UV, do not ionize and although fluorescent derivatives can be made, the procedure is tedious. Consequently, the RI detector can be ideal for detecting such materials and an example of such an application is shown in figure 18.

Figure 18

The Separation of Some Mono- and Di-Saccharides

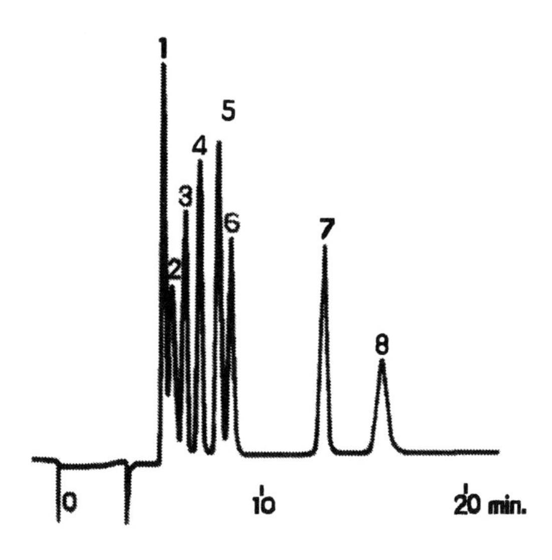

Courtesy of TOYO SODA Manufacturing Co. Ltd.

The separation was carried out on a TSKgel Amide-80 column 4.6 mm i.d. and 25 cm long with a mobile phase consisting of a 80% acetonitrile 20% water mixture. The flow rate was 1 ml/min and the column was operated at an elevated temperature of 80°C. The saccharides shown were 1/ rhamnose, 2/ fucose, 3/ xylose, 4/ fructose, 5/ mannose, 6/ glucose, 7/ sucrose and 8/ maltose. The analysis was completed in less than 20 minutes. These types of separations including other bio-monomers, dimers and polymers are frequently carried out employing refractive index detection.

Two further examples of similar types of analyses using the RI detector is afforded by the separation of the products of β–cyclodextrin hydrolysis and of the partial hydrolysis of galaction.

A chromatogram demonstrating the separation of the hydrolysis products of β–cyclodextrin is shown in figure 19.

Figure 19

The Separation of Hydrolyzed β–Cyclodextrin

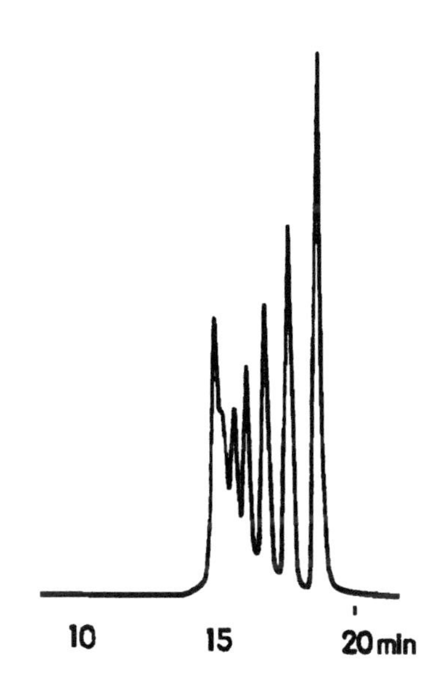

Courtesy of TOYO SODA Manufacturing Co. Lt

The separation was carried out on a TSKgel G-Oligo-PW column 7.8 mm i.d. and 30 cm long at 60°C and a flow rate of 1 ml/min. The TSKgel packing is a vinyl polymer based material suitable for separation by size exclusion using aqueous solvents. There are a number of grades of this material available that are suitable for separations covering a wide range of molecular weights. It is seen that the products of the hydrolysis are well separated and almost all of the oligomers are resolved.

The separation of the hydrolysates of galaction affords a further example of the use of the RI detector for organic polymers. The separation is again achieved by size exclusion and the chromatogram is shown in figure 20.

Figure 20

A Chromatogram of the Partial Hydrolysis of Galaction

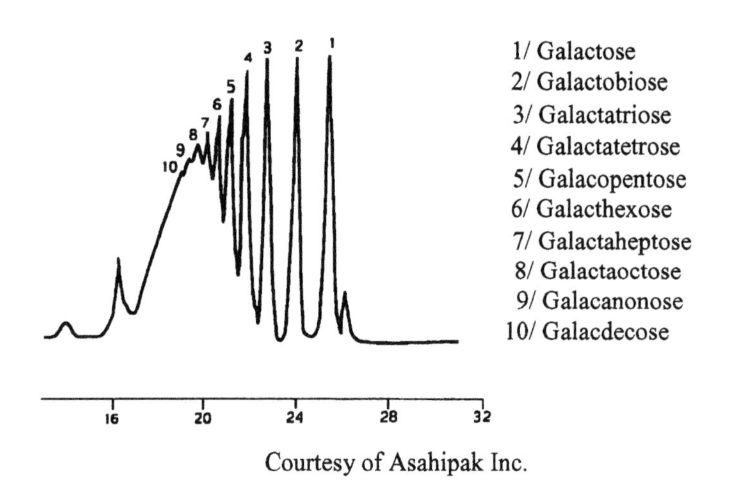

Courtesy of Asahipak Inc.

The columns used were a GS 220 x 2 50 cm long and 7.6 mm i.d.. The two long, wide columns were necessary as, in size exclusion chromatography, the total separation must be achieved in the pore volume of the column or about 40% of the dead volume. This problem has already been discussed and it is an interesting example of the practical use of large volume columns to provide adequate peak capacity. The mobile phase was, as is usual with many size exclusion separations, pure water. To improve the column efficiency and elevated column temperature of $60^{o}C$ was used with a mobile phase flow rate of 1 ml/min. For a column of 7.6 mm i.d. a flow rate of 1 ml/min would provide a linear mobile phase velocity close to the column optimum velocity. As discussed earlier, this provides the minimum HETP and thus, the maximum column efficiency. The conditions represent the result of a struggle to achieve adequate resolution by arranging for a large peak capacity to acommodate all the peaks (a large column volume) and an attempt to achieve the highest possible efficiency by operating close to the optimum velocity. Attempts were made to increase the column efficiency even further by reducing the dispersion due to the resistance to mass transfer in both phases by raising the column temperature to increase the solute diffusivity. Besides being an interesting example of the use of the refractive index detector, this separation is also an excellent example of how the operating condition of a chromatographic system can be adjusted to achieve a particular analysis.

The Tridet Mutifunctional Detector

The popularity of the UV detector, the electrical conductivity detector and the fluorescence detector motivated Schmidt and Scott (5,6) to develop a trifunctional detector that detected solutes by all three methods simultaneously in a single low volume cell.

Figure 21

The Trifunctional Detector

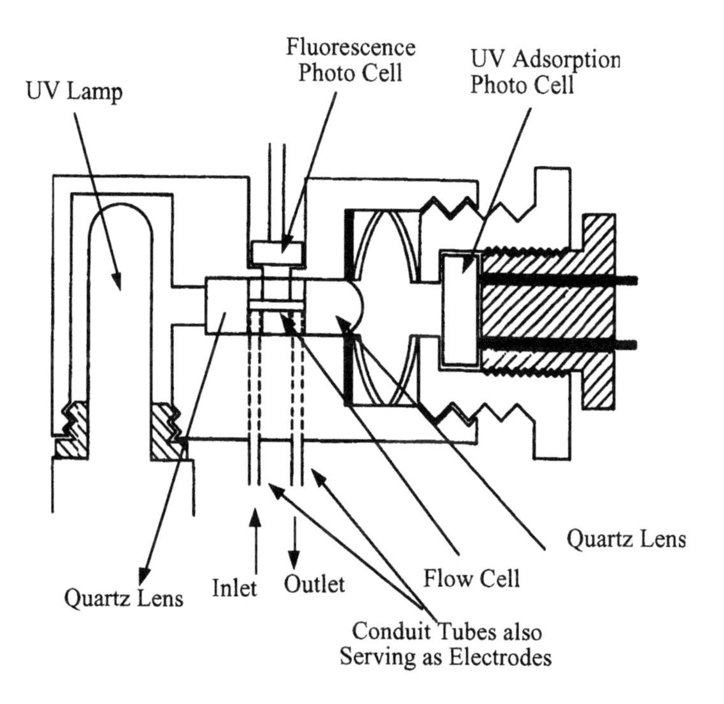

A diagram of their detector is shown in figure 21. The UV adsorption system consists of a low pressure mercury lamp emitting light at 254 nm and a solid state photo cell with quartz windows allowing the photo cell to respond to light in the UV region.

The detector cell is 3 mm long terminated at one end by a cylindrical quartz window and at the other by a quartz lens that focuses the transmitted light on to the photo cell. Next to the quartz windows are two stainless steel discs separated by a 3 mm cm length of Pyrex tube. The mobile phase enters and leaves the detector cell through radial holes in the stainless steel discs. The stainless steel discs also act as the electrodes for conductivity detection. At right angles to the Pyrex tube is situated another photo cell that receives any fluorescent light emitted normal to the UV excitation light. The output from each sensor passes to an appropriate amplifier to provide an output that is linearly related to solute concentration. Consequently, the column eluent is continuously and simultaneously monitored by UV adsorption, fluorescence and electrical conductivity. This detector, is obviously very versatile and, surprisingly, is relatively inexpensive. The detector together with a column, sample valve, pump and recorder, which constitutes a basic liquid chromatograph, costs about $8,000.

Chromatograms demonstrating the simultaneous use of all three detector functions are shown in figure 22. It is seen that the anthracene is clearly picked out from the mixture of aromatics by the fluorescence detector and the chloride ion, not shown at all by the UV adsorption or fluorescence detectors, clearly shown by the electrical conductivity detector.

However, it is not the simultaneous use of all detector functions that makes this detector so useful. The real advantage of the trifunctional detector is that it allows the analyst a choice of the three most useful detector functions in one detecting system. Furthermore, any of the three functions can be chosen at the touch of a switch and without any changes in hardware.

Figure 22

Chromatograms Demonstrating the Simultaneous Monitoring of a Mixture by All Three Detector Functions

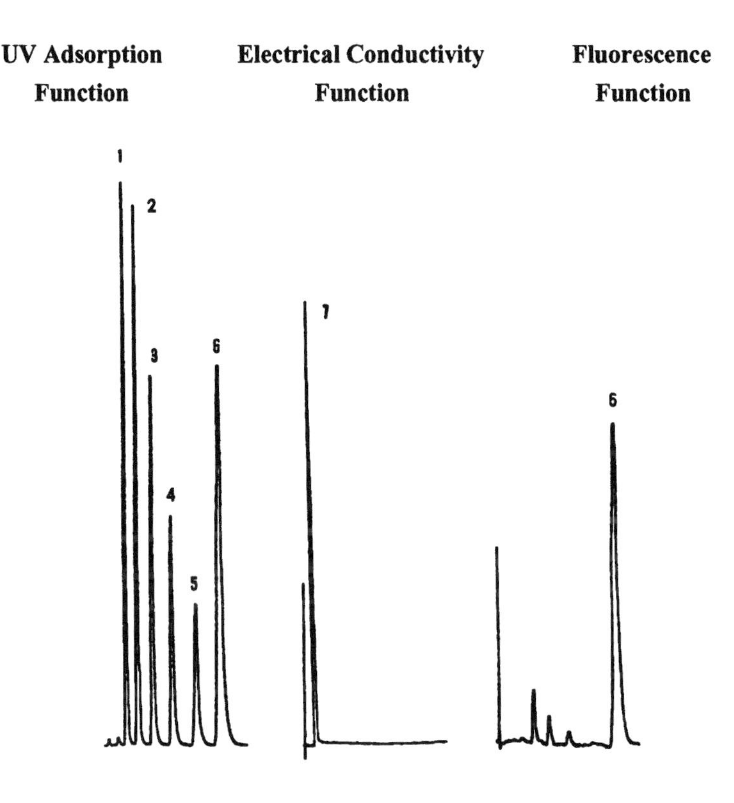

The column used was a Pecosphere™ 3 mm in diameter and 3 cm long carrying a C18 stationary phase. The mobile phase was a mixture of methanol (75%) and water (25%) at a flow rate of 2 ml/min. The solutes were 1 benzene, 2 toluene, 3 ethyl benzene, 4 isopropyl benzene, 5 t-butylbenzene, 6 anthracene, and 7 sodium chloride.

Courtesy of Bacharach Inc.

An example of the use of the three individual detector functions in the analyses of three quite different types of sample is shown in figure 23 and demonstrates this versatility.

Figure 23

Chromatograms Demonstrating the Different Functional Response of the TriDet Detector

Column Pecosphere
Size 3 mm x 3 cm C18
17% Methanol/Water

Flow Rate 3.0 ml/min

UV Detector

Column Pecosphere
Size 4.6 mm x 15 cm C18
90% Acetonitrile/Water

Flow Rate 2 ml/min

Fluorescence Detector

Column Pecosphere 3
Size 3 mm x 3 cm C18
1nMtetrabutyl-ammonium hydroxide and buffer
1.5 ml/min

Conductivity detector

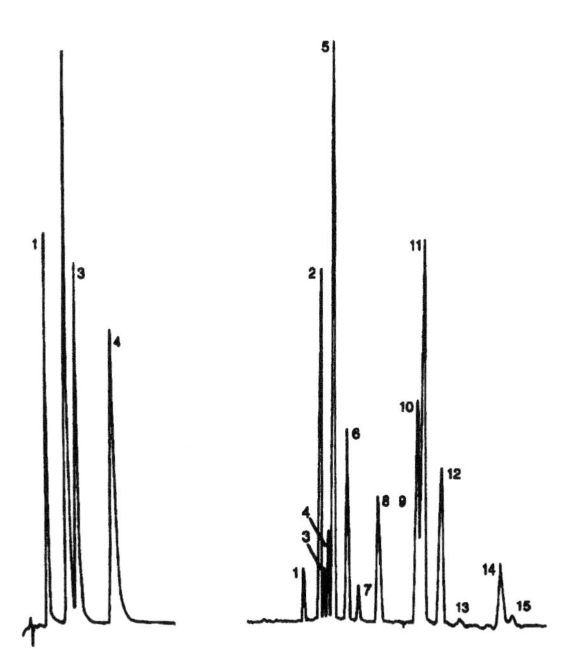

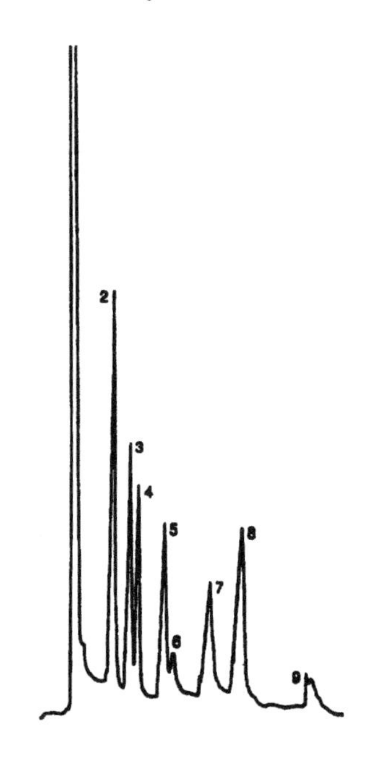

Sample Composition

1 Theobromine
2 Theophyline
3 Hydroxyethyltheophyline
4 Caffeine

1 Naphthalene
2 Fluorene
3 Acenaphthene
4 Phenanthrene
5 Anthracene
6 Fluoranthene
7 pyrene
8 Benzo(a)anthracene
9 Chrysene
10 Benzo(b) Fluoranthene
11 Benz0(k) Fluoranthene
12 Benzo(k) Fluoranthene
13Dibenz(a,h)anthracene
14 Idenol(1,2,3-cd))pyrene
15 Benzo(ghi)perylene

1 Solvent peak
2 Chloride ions
3 Nitrite ions
4 Bromide ions
5 Nitrate ions
6 Phosphate ions
7 Phosphite ions
8 Sulfate ions
9 Iodide ions

References

1/ "Liquid Chromatography Detectors" second edition, R.P.W.Scott, Journal of Chromatography Library, vol. 33, Elsevier, Amsterdam-Oxford-New York-Tokyo, (1986).

2/"Liquid Chromatography Detectors" (Ed. Thomas M. Vickrey) Chromatography Science Series,vol. 23, Marcel Dekker, Inc. New York (1983).

3/ I.A. Fowliss and R.P.W. Scott, *J. Chromatogr.* **11**(1963)1.

4/ E. Katz and R. P. W. Scott, *J. Chromatogr.*, **253**(1982)159.

5/ G.J. Schmidt and R.P.W. Scott *Analyst*, **110**(1985)757.

6/ G.J. Schmidt and R.P.W. Scott, U.S. Patent 4,555,936 (1983).

7
Sample Preparation

It is the intent of this chapter to introduce the analyst to some of the more common procedures that have been established for sample preparation. It is impossible to cover such a subject comprehensively in a single chapter and it will still be necessary for the analyst to seek support from the literature when faced with unusual samples. Fortunately, analytical LC methods have been reported in the literature for over two decades and it is highly likely that a publication exists describing a particular analysis of interest or one very similar to it. The journals that are recommended for reference are the *Journal of Liquid Chromatography,* the *Journal of Chromatography,* the *Journal of Chromatographic Science*, *The Analyst* and *Analytical Chemistry.*

The sample preparation in LC analysis is as important as the chromatographic separation itself. The procedure will often require considerable skill coupled with a basic understanding of chromatographic methodology. The analyst will need to have some familiarity with micro techniques including general micro-manipulation, micro-filtration, centrifugation and derivatization.

Sample Preparation Techniques

Sample preparation will involve extraction techniques, concentration procedures and methods for removing solid contaminants that may damage the sample valve, column or some other part of the chromatograph. The column can often be protected by the use of a pre-column or *guard column*, but such devices can introduce a significant extra column volume in the chromatographic system, causing further peak dispersion, and thus, impair resolution. If the components of the sample are well separated, then the extra dispersion resulting from the pre-column will not usually affect the separation adversely. If, however, the necessary separation is only just achieved with the phase system selected, then the sample dispersion that would occur in the guard column could well degrade the resolution to a point where its use could not be recommended. In the absence of a guard column, the column proper will only be protected by the cleaning procedure used in the sample preparation procedure prior to loading the column.

Filtration Techniques

After the sample has been prepared, any solid material that has carried through from the untreated sample, or has been introduced during the preparation procedures, should be removed before the sample can be safely loaded onto the column. Such precautions are necessary to protect the seats of the sample valve which are held in contact at very high pressures to prevent leaking. The most common method for cleaning the sample is by filtration. Traditional, hardened filter paper or a sintered glass filter can be used. However, the filter itself should be washed well with solvent particularly if filter paper is used to remove any lipid deposits that may have arisen from finger contact. If the materials of interest are present at a very low concentration, then a significant proportion of them may be adsorbed onto the solid material that is being removed from the sample. Alternatively a portion of the sample could be adsorbed on the surface of the filter medium itself. As the surface area of the filter is often fairly high, this loss can be substantial.

If the materials of interest are adsorbed on the surface of the solid impurities by *polar* forces, for example, due to contamination by substances such as silica, aluminum silicate or metal oxide, then they could be desorbed, by water, methanol, or acetonitrile if these solvents are present in the sample solution. Such solvents would also deactivate the cellulose of the filter paper so little adsorption would take place on its surface. However, if the materials of interest are adsorbed on the surface by *dispersive* interactions for example if the solid contaminants are organic or carbonaceous in nature, then the sample solution would need to contain a solvent that was more *dispersive* such as tetrahydrofuran or perhaps an appropriate aliphatic ester. When such a solvent modification is being considered, then the effect of foreign solvents on the selectivity of the column must also be borne in mind. However, if the sample volume is less than 5 μl, and a small bore column (i.d. 1 mm or less) is *not* being used, the column selectivity should not be significantly affected by the composition of the sample solvent mixture.

Special solvents that are not components of the mobile phase, but are included in the sample to improve component solubility, will act as though they were solutes themselves. Each will produce a spurious peak somewhere on the chromatogram that must not be misinterpreted as a solute peak. Irrespective of the sample solvent, the solutes of interest must always be sufficiently soluble in the mobile phase to permit effective chromatographic development.

A general method of eliminating the adsorption of trace materials on solid residues, or container walls, is difficult to predict and each situation must be evaluated on the basis of its own unique conditions. It should be emphasized that, assuming there are only small quantities of contaminating materials present that could adsorb a proportion of the sample, significant losses would only arise when the solutes of interest are present in the sample at *the ppm level*. Such problems of adsorption are reduced considerably if the contaminating solid material is removed by centrifugation.

Centrifugation Techniques

Removal of solid debris using a micro-centrifuge is an alternative technique to filtration. Although it requires special apparatus, the process has much to commend it. Firstly there is no filter medium on which trace components of the sample can be adsorbed, secondly and more importantly, the sample volume is not changed during solid removal, and no sample is lost. Micro-centrifuges are particularly valuable for removing cell debris from biological or environmental samples. Small glass micro-centrifuge tubes can be used but, if strongly polar trace-materials are present, they may be adsorbed by polar forces on to the hydroxyl groups on the wall of the glass tube; under such circumstances treated tubes should be used.

The tube wall can be deactivated by reacting the surface hydroxyl groups with an appropriate alkyl silane. This attaches a hydrocarbon chain to the hydroxyl groups rendering the wall highly dispersive in character. As a consequence, the polar solutes do not interact strongly with the glass wall and there is minimal loss by adsorption. Micro-centrifuge tubes, already treated with a silane reagent, are commercially available. It must be stressed, however, that if the sample contains trace materials that are significantly *dispersive* in character, then the converse applies and they will be rapidly adsorbed on the surface of the silanized glass wall. Under such circumstances it is important to use *untreated* centrifuge tubes for such samples.

It is clear that the nature of the sample and the nature of the materials with which the sample comes in contact are very important considerations in trace analysis. Correct choices can be made easily, on a rational basis, providing the nature of the molecular interactions that can take place are known and understood.

Concentration and Extraction Techniques

As a result of limited detector sensitivity, there is often a need for sample concentration when determining trace materials contained in a bulk matrix. The need for such procedures frequently arises in forensic work, environmental samples, blood testing etc. A number of methods have been developed for this purpose and some of those in common use will be described.

Evaporation

In the isolation of trace compounds from bulk material, the substances of interest are often extracted or recovered in a volatile solvent. The solutes can be concentrated by direct evaporation, providing the solutes are *sufficiently non-volatile*. In fact, all solids have a finite vapor pressure, some quite large (e. g. naphthalene) and, as the compounds are present in very low concentration, they need only have a very low, but significant vapor pressure for substantial quantities to be lost in the evaporation process. In cases where the substances in solution do have a sufficiently low vapor pressure, their concentration can be increased by removing the solvent in a stream of nitrogen or argon. Air cannot be used as the oxygen present is likely to oxidize some, or all, of the components. In the evaporation process, the solution is contained in an appropriate glass tube and care must be taken to use a *slow* stream of the gas flowing over the surface of the solvent to prevent loss by splashing. Sometimes, if the vapor pressure of the solvent is relatively low it can be increased by raising the temperature of the sample to 40-50°C, but care must be taken with thermally labile substances. Evaporation should be continued to reduce the volume of the solvent below that actually required and then made up accurately to specific a volume to assure quantitative accuracy.

Lyophilization

Lyophilization is a similar technique and is, in fact, evaporation at reduced temperature under vacuum. In some cases, an aqueous sample can be frozen and the vapor pressure of the ice is sufficient to produce a relatively rapid rate of evaporation. It can also be used effectively where the substances of interest have vapor pressures that are sufficiently high at room temperature, to cause substance loss under normal evaporation procedures. As the vapor pressure of a substance is exponentially related to the temperature, a relatively small reduction in temperature can reduce the vapor pressure of the sample components sufficiently to render any loss during evaporation relatively insignificant. This technique is gentler than evaporation and, for this reason, is often used for samples of biological origin that are labile, and for substances such as proteins, which denature easily. Specific equipment is necessary to lyophilize a solution, but such apparatus is readily available.

Precipitation

Precipitation is another method of concentration that is used extensively for biopolymers such as proteins, polypeptides, etc. By increasing the salt concentration of a protein solution, some proteins can be precipitated. The sample is then centrifuged and the supernatent liquid removed. The protein can be reconstituted in an appropriate solution, usually the one that will subsequently be used as the mobile phase.

This precipitation process can be carried out rather cleverly on the surface of a reverse phase. If the protein solution is brought into contact with a reversed phase, and the protein has dispersive groups that allow dispersive interactions with the bonded phase, a layer of protein will be adsorbed onto the surface. This is similar to the adsorption of a long chain alcohol on the surface of a reverse phase according to the Langmuir Adsorption Isotherm which has been discussed in an earlier chapter. Now the surface will be covered by a relatively small amount of protein. If, however, the salt concentration is now increased, then the protein already on the surface acts as deposition or 'seeding' sites for the rest of the protein. Removal of the reverse phase will separate the protein from the bulk matrix and the original protein can be recovered from the reverse phase by a separate procedure.

Solid State Extraction Cartridges

The 'solid phase extraction cartridge' (SPEC) is another somewhat vainglorious name given to a short inert plastic tube packed with an adsorbent, usually a reversed phase or an ion exchange resin. The particle size of the packing is often significantly larger than that used in LC columns to ensure a reasonable permeability. A diagram of a solid state extraction cartridge is shown in figure 1.

Figure 1

A Solid State Extraction Cartridge

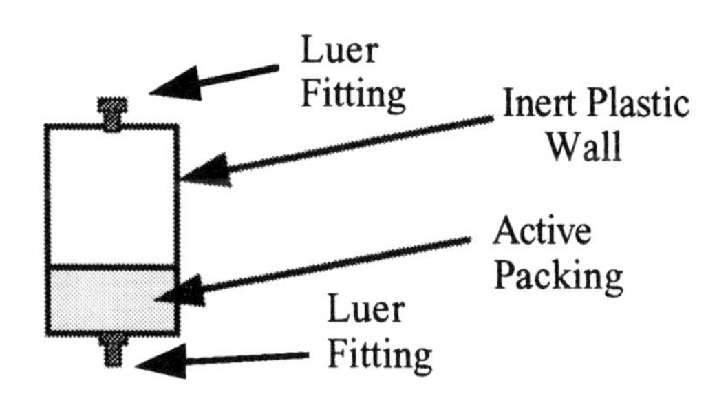

It is seen that the plastic tube can be connected to a hypodermic needle by means of a Luer fitting if so desired or, alternatively, the top of the cartridge can be connected to a syringe to provide solvent flow. The cartridges are available in a range of sizes, the 1, 2 and 5 ml being the most popular. The solution is poured into the top of the tube and allowed to percolate through the packing. Devices are available to allow a vacuum to be applied to the base of the tube or a pressure to the top. By choosing the appropriate adsorbent, which may range from a silica gel to an affinity packing, a wide variety of materials can be selectively adsorbed on to the surface. The adsorbent is usually prepared by washing with an appropriate solvent mixture and then the sample solution is allowed to percolate through it so that the material of interest is adsorbed from the sample matrix. The adsorbent is then washed free of the matrix liquid with another solvent and finally desorbed into yet a third solvent. Thus, the material from as much as a 100 ml of solution can be extracted and concentrated and by desorbing the adsorbed components into 1 ml of solvent a concentration factor of 100 would be achieved. The following examples serve to illustrate the use of extraction cartridges.

The tetrahydrocannabinol carboxylic acid was extracted from the urine by means of a solid state extraction cartridge packed with a C18 reverse phase (octyldecyldimethyl chains). As the urine sample was used direct, and contained no added solvent, the materials of interest were irreversibly adsorbed on the reverse phase solely by dispersive interactions.

The procedure for extracting tetrahydrocannabinol carboxylic acid was as follows (1): 5 ml of urine was mixed with 5 ml of water and 0.5 ml of 10M potassium hydroxide and placed in a silanized glass tube. Note the tube with a non-polar wall was used to ensure that none of the acid was adsorbed on the hydroxyl groups of the glass. The tube was heated to 60°C for fifteen minutes and then cooled to room temperature. The quantity of silane remaining on the glass surface after heating with alkali for fifteen minutes is an interesting point and raises the question as to whether a treated tube was needed in the first place.

The sample was then adjusted to a pH of 4.5 with acetic acid. The C18 reversed-phase in the extraction tube was pre-conditioned with a mixture of 2 ml of methanol and 1 ml of 1% acetic acid. The hydrolyzed sample was allowed to percolate slowly through the tube by dropwise addition. The tube packing was then washed twice with 1 ml of a 20% aqueous solution of acetone, once with 1.5 ml aqueous 0.01M KH_2PO_4 , once with 0.5 ml of aqueous 0.01M Na_2HPO_4 and finally 0.5 ml of 1% aqueous acetic acid. The sample was then slowly desorbed from the reversed-phase with 1 ml of pure methanol and collected in a silanized glass tube. The sample was reduced in volume by evaporation in a stream of nitrogen and finally made up to a volume of 250µl with methanol. The chromatograms obtained from the urine sample and a reference standard is shown in figure 2. It is seen from figure 2 that the tetrahydrocannabinol carboxylic acid is clearly and unambiguously separated from the contaminating materials with an extraction efficiency of over 90%. This is a typical application for solid phase extraction cartridges.

Figure 2

Chromatograms of Tetrahydrocannabinol Carboxylic Acid from Urine

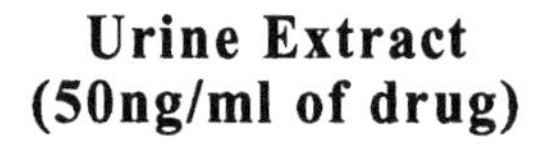

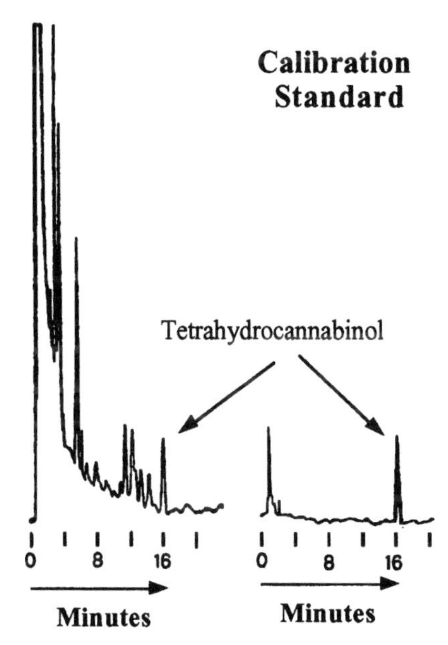

Courtesy of Supelco Inc.

Chromatographic Conditions

Column	LC 18
Column Length	25 cm
Column Diameter	4.6 mm
Column Packing	C18 Reverse Phase (octadecyldimethyl chain)
Mobile Phase	55% acetonitrile 45% aqueous 1% acetic acid
Flow Rate	2.5 ml/min
Detector	UV adsorption at 280 nm
Sample Volume	100μl containing 500 ng of tetrahydrocannabinol

The Determination of the Tricyclic Antidepressant Drugs from Serum

Another interesting example of the use of solid state extraction cartridges is the determination of the tricyclic antidepressant drugs in blood serum (2).

The solid phase used in this extraction is a weak ion exchanger and the material is preconditioned with a wash of 0.5 ml of 0.5M phosphoric acid followed by 1 ml of deionized water. A volume of 0.5 ml of the serum containing the tricyclic antidepressant drug standards is mixed with 0.5 ml of deionized water and allowed to percolate slowly through the packing.

The drugs are held on the ion exchanger whereas the sample matrix materials pass through. The packing is then washed with 0.5 ml of 1.0M aqueous ammonium hydroxide and then two, 1 ml aliquots of 5% methanol in water. The sample is then desorbed by two separate aliquots of 1 ml of 0.22M ammonium hydroxide in pure methanol. Finally, the sample is collected in a silanized glass tube and the solvent removed by evaporation under stream of dry nitrogen. The volume of the sample is then made up to 250 μl and 100 μl placed on the column. The separation obtained is shown in figure 3.

Chromatographic Conditions

Column	LC-PCN
Column Length	25 cm
Column Diameter	4.6 mm
Column Packing	C18 Reverse Phase (cyanopropile chain)
Column Temperature	30°C
Mobile Phase	25%0.01M potassium phosphate (adjusted to pH 7 with 85% phosphoric acid) 60% acetonitrile and 15% methanol
Flow Rate	2 ml/min
Detector	UV adsorption at 215 nm
Sample Volume	100μl

Figure 3

Chromatogram of the Tricyclic Antidepressant Drugs from Blood Serum

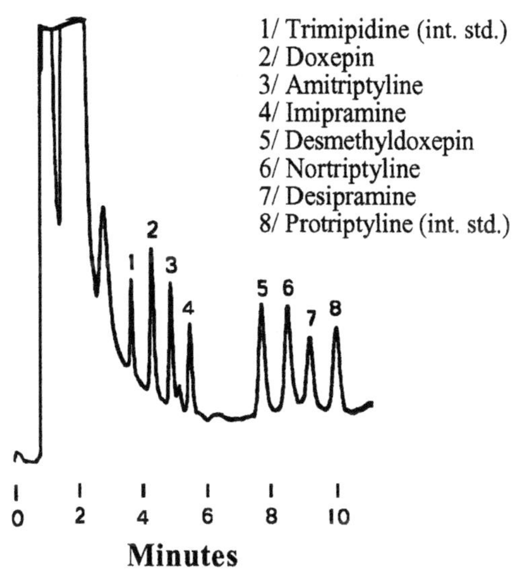

Courtesy of Supelco Inc.

The separation was carried out on a bonded phase LC-PCN column carrying cyanopropylmethyl moieties on the surface. Thus, in contrast to the extraction process, which appears to be based on ionic interactions with the weak ion exchange material, the LC separation appears to be based on a mixture of interactions. There will be dispersive interactions of the drugs with the hydrocarbon chains of the bonded moiety and also weakly polar interactions with the cyano group. It is seen that the extraction procedures are very efficient and all the tricyclic antidepressant drugs are eluted discretely.

An Automatic Sample Extraction and Concentration Procedure

An automatic solid phase extraction system was developed by Scott and Kucera (3) for the determination of drug metabolites in blood. The system was basically a multi-column switching system utilizing solid phase extraction and the apparatus used is shown in figure 4. The sampling device was designed for use with small bore columns (4) as they could provide the high mass sensitivities necessary for measuring low concentrations of metabolites in blood and blood serum. The apparatus consisted of two six-port Valco valves that would function up to pressures of 7000 lbs/sq in. The valves were used in conjunction with a small single piston Eldex pump. The first valve controls an open sample loop, the volume of which can be chosen to suit the sample.

Figure 4

Apparatus for Automatic Sample Extraction and Concentration

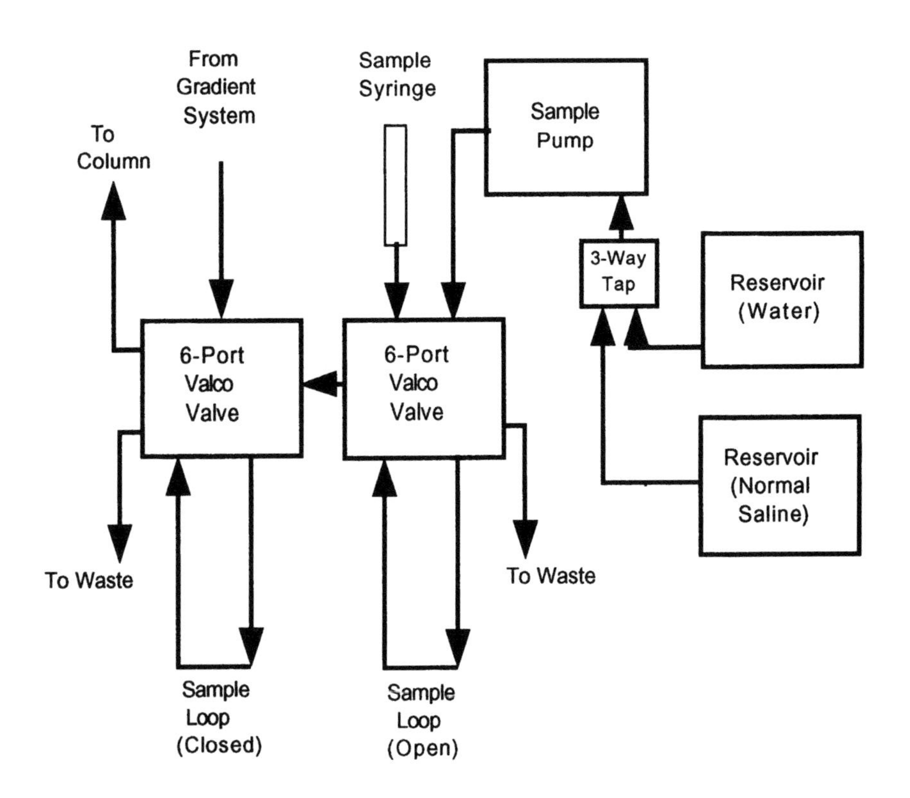

The second valve controls a sample loop, 5 cm long and 1 mm in diameter, packed with dimethyloctadecyl reverse phase comprising of fairly coarse particles 100-120 μm in diameter to reduce flow impedance. The sample pump is supplied via a two-way tap from either of two reservoirs, one containing pure water and the other, 'normal' saline. The output of the pump can be used to either force the contents of the open loop sample tube through the packed loop, or to permit washing with an appropriate solvent. The separate pump is necessary to overcome the impedance of the packed loop.

The sampling procedure is as follows. The system comprising the two sample loops is first filled with saline to prevent protein being precipitated from the serum which can occur if it comes in contact with pure water. A sample syringe is used to fill the open loop with the sample of blood serum to be analyzed. The sample pump is then activated and saline is used to force the serum sample through the packed tube and sufficient saline is used to displace all the extracted serum to waste. The drug metabolites and any other materials that carry dispersive groups will interact strongly with the reversed-phase and be held tightly to the packing. The saline is now replaced by pure water and the saline washed from the system. The sample now resides on the front part of the reversed-phase packing. The second valve is now rotated and the packed loop placed in line with the mobile phase supply to the column. The gradient is now commenced and the separation developed. It should be noted that the mobile phase enters the loop from the opposite end to that which the sample entered and thus, the adsorbed material is desorbed by the mobile phase directly onto the LC column. In this way the sample is not required to traverse the packed trap before entering the column. After the sample has been placed on the column, the valves are returned to their primary position and the system filled with saline in readiness for the next analysis.

The overall sensitivity of the sampling and concentrating system was examined using acetophenone as the solute. Purified water (water that had been passed through a reversed-phase column) was used to make up standard solutions containing 1 ppm, 100 ppb, 10 ppb, and 1 ppb. For these experiments the saline wash was superfluous. The samples of acetophenone solutions were drawn into the open loop, the contents displaced by pure water from the pump through the packed loop and the sample desorbed on to the column in the manner previously described. The column was 50 cm long, 1 mm in diameter and packed with ODS 2 reversed phase having particles 10 µm in diameter. The mobile phase was a mixture of 75% methanol and 25% water and the column was operated at a flow rate of 40 µml/min. The results obtained are shown in figure 5.

Figure 5

Chromatograms Demonstrating the Concentration Effect of the Microbore Column Sampling Apparatus

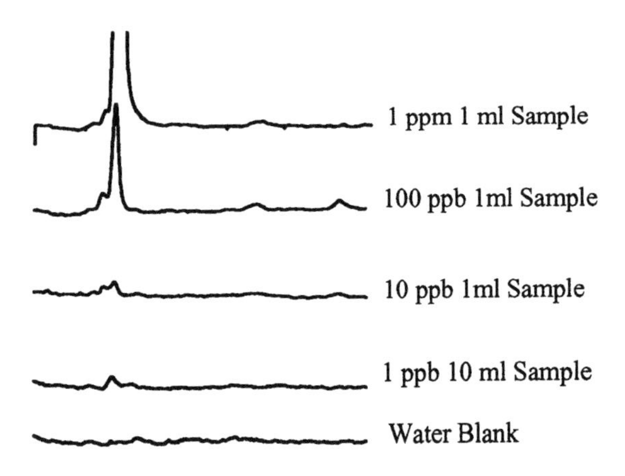

It is seen that with 1 ml samples the peak for a concentration of 1 ppm is well off scale and a clearly defined peak was observed for 10 ppb. When a 10 ml sample was used, acetophenone at a concentration of 1 ppb could just be detected. Under the conditions used, the ultimate mass sensitivity of the system was about 10 ng. This level of sensitivity was achieved, on the one hand, by the sample concentrating process, and on the other, as a result of the high mass sensitivity of small bore columns.

The column used for blood serum analysis was 100 cm long, 1 mm in diameter and packed with RP 18 reversed phase having a particle size of 10 μm. A concave gradient program was used to develop the separation over a period of 45 min. at a flow rate of 50 μl/min. The initial solvent was 75% methanol 25% water and the final solvent was pure methanol.

A sample loop, having a capacity of 400 μl, was filled with the serum which was displaced by 2 ml of normal saline onto the packed sample loop. The loop was then washed with a further 2 ml of saline followed by 2 ml of water and the placed in line with the solvent flow and the program initiated. The chromatogram obtained is shown in figure 6.

Figure 6

Chromatograms of Blood Serum Showing Dispersive Contaminants at the ppm Level

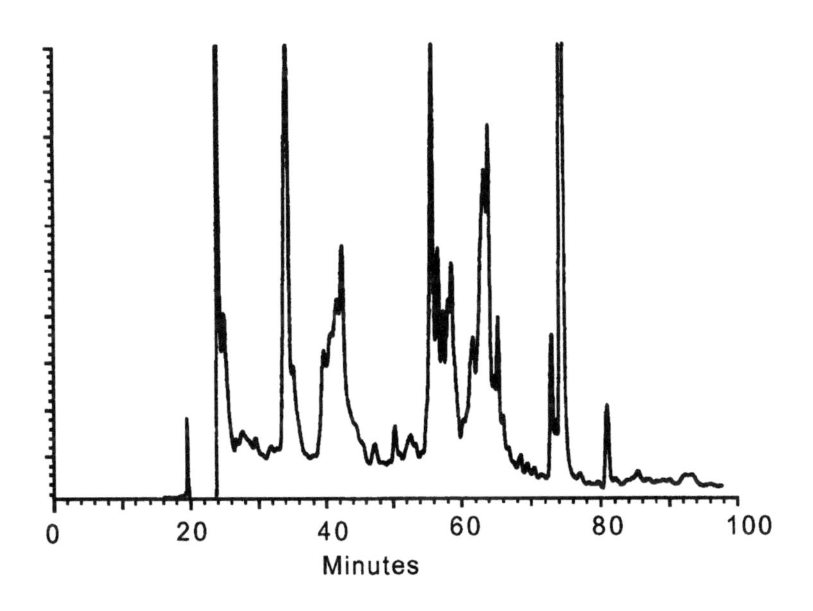

It is seen that there are a large number of compounds capable of dispersive interactions with the reverse phase, contained in the serum that have been extracted and separated. Again the results have been obtained, partly as a result of the extraction and concentration properties of the sampling system, and partly as a result of the high mass sensitivity of the small bore columns.

Solid phase extraction systems have a wide range of application whether operated manually or automatically. Wherever materials are present in solution at very low concentrations they offer a means of extraction and concentration that is simple, efficient and fast.

A Sample Separation Protocol

The methodology involved in sample preparation must achieve a number of objectives.

1/ A representative sample must be placed on the column to ensure accurate quantitative results.

2/ The sample solvent or other constituents must neither affect the column performance, temporarily or permanently, or the column resolution.

3/ The sample must not contaminate the sample valve or the column with materials that may cause damage.

4/ The detector must respond to the samples of interest and if not, then the necessary derivatization procedures must be carried out.

5/ The concentration of the solutes of interest must be such that an adequate response is given by the detector.

The sample preparation discussed will not include the addition of standards for quantitative analysis as this subject will be dealt with in the chapter on quantitative analysis. Samples can arrive for LC analyses as solids, liquids or a mixture of both and, therefore, the three possibilities must be considered separately. Furthermore, the components of interest may be present at high or low concentrations and consequently, will require either concentration or dilution for effective LC analysis. Thus, there are six broadly different procedures that need to be discussed.

Before the actual sample preparation procedure is described some general observations should first be made. However excellent the sample preparation and however sophisticated the equipment, the accuracy of the analysis will only be as good as the quality of the sample that is taken. If the sample is that of a reaction mixture from an organic synthesis laboratory, it is likely to be taken from a single bottle or container, by a professional chemist, and is likely to be truly representative of the bulk of the material.

At the other extreme, if the sample is of river water and, for example, is to be analyzed for pesticides, the representative quality of the sample becomes far more uncertain. However carefully the sample is extracted and concentrated and prepared for analysis, if the sample had been taken after it had rained all night, or two days after the neighboring farmer had sprayed his fields with pesticides, the results will not be reliably representative of the river contamination. There are correct ways of taking such samples and as part of the responsibility of sample preparation, sampling procedures must be ascertained and evaluated. The sampling method should be reported with the analytical data to allow the significance of the results to be estimated. This is particularly important in LC analysis as the technique is extremely sensitive and is frequently used for the analysis of trace materials in environmental analysis, drug assays, medical tests, forensic samples etc., where the results are frequently used for litigation purposes. Sampling methods are, themselves, complex procedures but their discussion is outside the scope of this book. Those interested in pursuing the subject further should refer to the appropriate publications of the A.S.T.M. (the American Society for Testing Materials).

In sample preparation, it is assumed that the type of phase system that will be used for the separation has been established using pure samples of the solutes of interest. Under some circumstances the phase system for the separation may be developed by using the actual sample for analysis but this is rare. In any event, whether the correct phase system has been identified using pure reference substances, or the sample itself, it will have been established prior to the analysis proper. Thus, by the time the analysis is to be carried out, the column will have been chosen and the composition of the mobile phase will be known. A knowledge of the components of the mobile phase will provide help in choosing an extraction solvent or dilution solvent should it be required. Ideally the extraction or dilution solvent should be the mobile phase itself, or, if the substances are insufficiently soluble in the mobile phase mixture, then the component of the mobile phase that has the greatest impact on the elution rate would be the first choice. For example if the mobile phase was a mixture of acetonitrile and water, and the materials were not sufficiently soluble, then either a solvent mixture containing more acetonitrile should be tried or at the extreme, pure acetonitrile itself.

A number of different sample preparation procedures will now be described to illustrate how the appropriate method will vary, both with the physical nature of the sample, and the chemical character of the components of interest. The examples have been taken from a variety of sources, including application notes from the manufacturers of stationary phases and different chromatography journals.

LC Sample Preparation for Solid Materials

Solid samples are generally treated in one of two ways. If completely soluble, they can be dissolved directly and completely in a suitable solvent. Alternatively, if the samples contain insoluble materials that are of no interest, then they can be extracted with a selected solvent to obtain the relevant compounds in solution. The extract can be subsequently filtered or centrifuged to remove any unwanted substances that make up the sample matrix. The procedure will differ, depending on the amount of the substances present that are germane to the analysis. The preparation of samples for LC analysis from solid materials will be considered both when the solutes of interest are present at a high concentration and also when present in trace quantities.

Materials of Interest Present in High Concentration

Typical examples that fall in this group would be the determination of the active ingredients in analgesic tablets for pharmaceutical use, such as aspirin or codeine or the analysis of a food product such as margarine. Examples of both these analyses will be described to illustrate the sample preparation procedure.

The Determination of Free Fatty Acids in Margarine (5)

Margarine is an example of a solid sample where the materials of interest are soluble in one solvent (in this case methanol) whereas the matrix materials, largely triglycerides, are not. As a consequence, the sample preparation procedure is relatively simple. The chromatographic separation is achieved by using the dispersive interactions between the hydrocarbon chains of the fatty acids and the hydrocarbon chains of a reversed phase.

In general a known mass of the sample is dispersed in a solvent in which the substances of interest are known to be soluble. The amount taken will depend on the amount of material in the sample, the volume of charge that will be placed on the column and the sensitivity of the detector. As a guide to the sample volume, if a small bore column is used (1 mm i.d.) or a short column packed with particles 2-3 μm diameter, then the peaks may be only a few microliters wide. Consequently, the maximum volume charge is likely to be less than 2 μl. If larger columns are used or longer columns packed with larger particles, sample volumes of up to 100 μl or more may be acceptable without serious band dispersion. A weighed quantity of margarine (about 1 g) was added to 5 ml of methanol, warmed to melt the margarine and well shaken. Fatty acids dissolve in the methanol whereas the triglycerides do not. The mixture was allowed to settle and 1 ml of the methanol layer was taken and diluted in a standard flask. A portion of the diluted sample was filtered, and 10 μl of the filtered sample which, in the example given, contains about 10 μg or free fatty acids, was placed on the column and the separation developed isocratically. The chromatogram obtained is shown in figure 7.

Figure 7

Chromatogram of Free Fatty Acids from Margarine

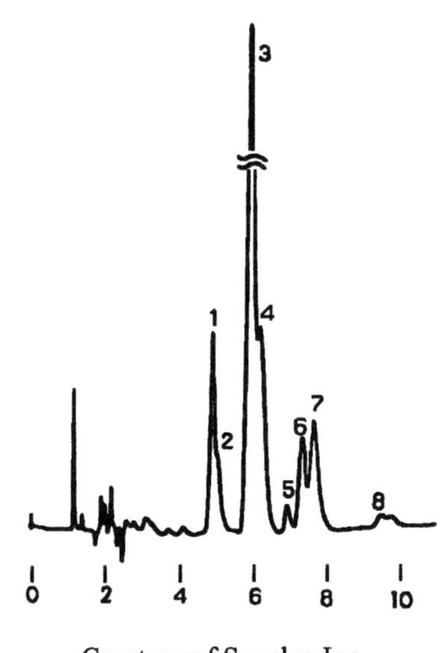

Courtesy of Supelco Inc.

The solutes labeled in the chromatogram are as follows:

Peak Number	Solute
1	Linolenic Acid
2	Myristic Acid
3	cis-Linoleic Acid
4	trans-Linoleic Acid
5	Palmitic Acid
6	cis-Oleic Acid
7	trans-Oleic Acid
8	Stearic Acid

Chromatographic Conditions

Column	LC 8
Column Length	25 cm
Column Diameter	4.6 mm
Column Packing	C8 Reversed Phase (octyldimethyl chain)
Mobile Phase	54% acetonitrile 21.5% tetrahydrofuran 28% of 0.1 % phosphoric acid in water
Flow Rate	1.6 ml/min
Detector	UV adsorption at 215 nm
Sample Volume	10 μl

As reverse phase column (LC-8) was used (octyldimethyl chains), the interactions with the stationary phase that controlled retention, and thus, the separation, were predominantly dispersive. It should be remembered, however, that although dispersive interactions dominate in the stationary phase, the separation is controlled by adjusting the competing dispersive interactions together with the polar interactions that take place in the mobile phase by the solvent composition.

The Analysis of Acetaminophen and Aspirin Tablets

The analysis of a pharmaceutical tablet (6) requires sample preparation that is little more complex as most tablets contain excipients (a solid diluent) that may be starch, chalk, silica gel, cellulose or some other physiologically inert material. This sample preparation procedure depends on the insolubility of the excipient in methanol. As the components of interest are both acidic and neutral, the separation was achieved by exploiting both the ionic interactions between the organic acids and the adsorbed ion exchanger and the dispersive interactions with the remaining exposed reverse phase.

To remove the excipient, the tablet was ground to a powder and a weighed portion treated with a known volume of a mixture of 5% glacial acetic acid in methanol. The slurry was well stirred to ensure all the active ingredients were dissolved and the mixture was filtered.

A known volume of the filtrate was taken and quantitatively diluted to a volume appropriate for the sample volume that was to be used and the sensitivity of the detector that was to monitor the separation. The chromatogram obtained is shown in figure 8.

Figure 8

Chromatogram of the Active Ingredients of Acetaminophen and Aspirin Tablets

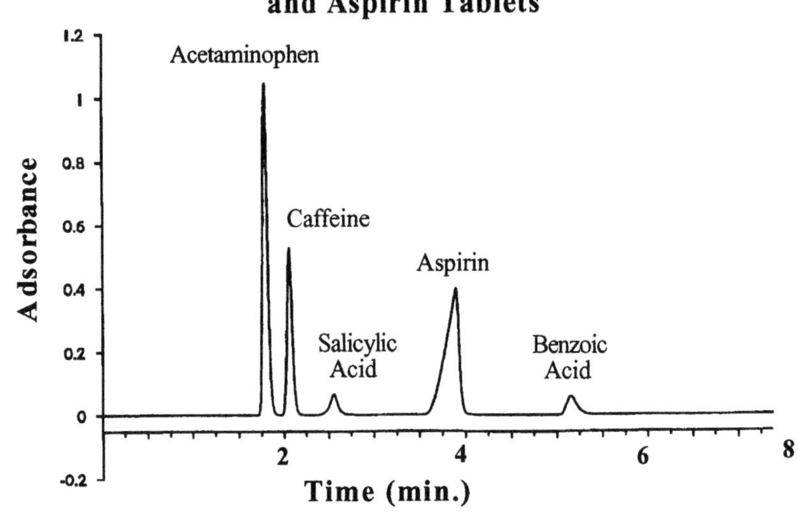

Courtesy of Whatman Inc.

Chromatographic Conditions

Column	C18 (5 µm particle size, pore size 300 Å)
Column Length	25 cm
Column Diameter	4.6 mm
Column Packing	C18 Reverse Phase (octyldecyldimethyl chain)
Mobile Phase	75%water,12%methanol 12%acetonitrile and 1%acetic acid and 225 mg of tetramethylammonium hydroxide per l
Flow Rate	1.6 ml/min
Detector	UV adsorption at 254 nm

As the solvent mixture also contained 225 mg of tetramethyl ammonium hydroxide pentahydrate per liter at a high water content (75%), the surface of the reverse phase would have been largely covered with the tetramethyl ammonium hydroxide pentahydrate. This would have acted as an adsorbed ion exchange stationary phase. It is clear that the free acids, salicylic acid, acetylsalicylic acid (aspirin) and benzoic acid were retained largely by ionic interactions with adsorbed basic ion exchanger and partly by dispersive interactions with the exposed reversed phase. The acetaminophen and the caffeine, on the other hand, being unionized substances, were retained only by dispersive interactions with the exposed reversed phase.

Materials of Interest Present in Low Concentration

The two examples of sample preparation of solids containing low concentrations of the substances of interest will be the analysis of aflatoxins in corn meal (7) and the determination of the fungicide thiabendazole in citrus fruits. It should be pointed out that the applications chosen in this section attempt to reflect a range of analyses that the analyst is likely to meet in both research and industrial laboratories.

The Analysis of Aflatoxins in Corn Meal

This sample preparation involved, firstly, an extraction and the elimination of the solid matrix by filtration and, secondly, a concentration procedure employing a solid phase extraction cartridge. The compounds of interest were separated solely by dispersive interactions with the reversed phase. In the example given, the corn meal was 'spiked' with the aflatoxins.

Fifty grams of the corn meal was blended with 100 ml of an 80% methanol 20% water mixture for about a minute and the corn meal removed by filtration. The extraction cartridge was packed with a bonded phase carrying cyanopropylmethyl chains that would exhibit both weakly polar and dispersive interactions with the aflatoxins. The cartridge was preconditioned with 2 ml of 0.5% acetic acid in water.

One milliliter of the filtered corn meal extract was mixed with 4 ml of the 0.5 % acetic acid solution and allowed to percolate through the cartridge.

The packing wash consisted of 0.5 ml 20% of tetrahydrofuran in a 0.5% solution of acetic acid in water, followed by 2 ml of hexane. The packing was then dried in a stream of nitrogen and washed with a further 3 ml of 25% tetrahydofuran in hexane and again dried under nitrogen. The recovered aflatoxins were finally eluted with two 2 ml aliquots of 1% tetrahydrofuran in methylene dichloride into a silanized tube.

Figure 9

Chromatogram Showing the Aflotoxins in Corn Meal

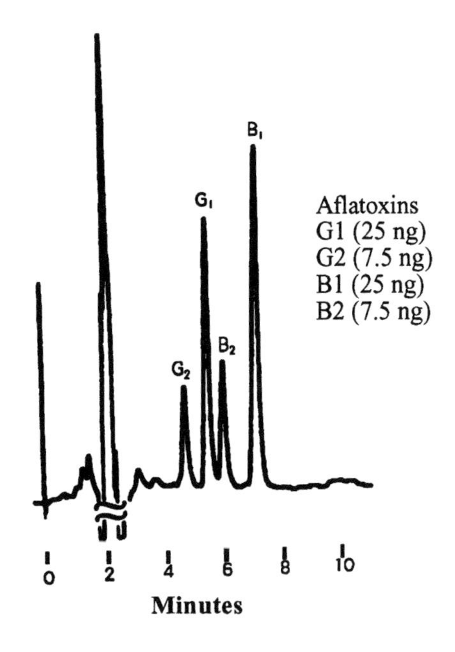

Courtesy of Supelco Inc.

Chromatographic Conditions

Column	LC 18
Column Length	25 cm
Column Diameter	4.6 mm
Column Temperature	Ambient
Column Packing	C18 Reversed Phase (octydecylldimethyl chain)
Mobile Phase	52.5% acetonitrile 22.5% methanol 55%water
Flow Rate	1.5 ml/min
Detector	UV adsorption at 365 nm
Sample Volume	100 μl

Just prior to analysis the contents of the tube were evaporated to dryness. The residue was then re-dissolved in 100 μl of methanol and then diluted with 100 μl of 0.5 % acetic acid in water. 100 μl of the resulting mixture solution was then placed on the column and the chromatogram obtained is shown in figure 9. It is seen that an excellent separation was obtained and again the retention and selectivity was dominated by dispersive interactions with the reversed phase.

The sample preparation is generally typical of the LC analysis of many foodstuffs although the specific substances of interest will differ widely.

The Determination of the Fungicide Thiabendazole in Citrus Fruits (8)

This sample preparation required both extraction and concentration and this was carried out in the traditional manner with the use of an appropriate solvent. The separation was again achieved exploiting the dispersive interactions between the components of the mixture and the strongly dispersive hydrocarbon chains on the reversed phase.

Between 5 and 10 individual fruits were each sliced into 16 parts. Randomly 20 to 40 parts were selected and homogenized in a juicer. 10 gm of the mixture was weighed into a blender with 3 g of anhydrous disodium phosphate, 10 g of anhydrous sodium sulfate and

80 ml of ethyl acetate and homogenized for 10 min. The supernatent liquid was filtered into a 300 ml flask. The precipitate was homogenized with two further aliquots of 80 ml of ethyl acetate and the mixture of the combined filtrates was then concentrated under vacuum to about 5 ml on a water bath (40 °C). The liquid was then diluted to 20 ml with methanol in a standard flask, filtered once again and 10 µl placed on the column. The results obtained are shown in figure 10.

Figure 10

Chromatogram of Thiabendazole Extracted from Citrus Fruits

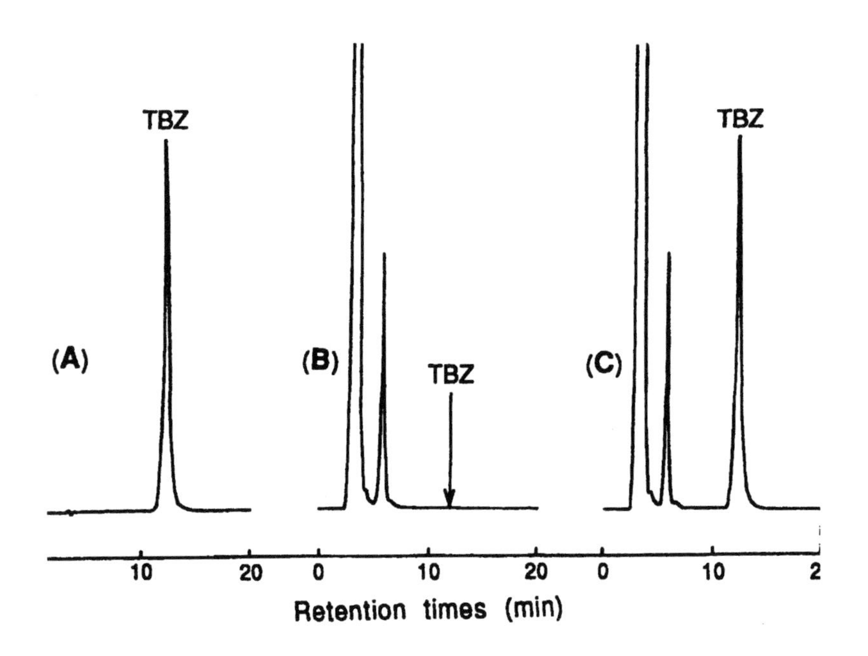

Chromatographic Conditions

Column	Unicil QC8
Column Length	5 cm
Column Diameter	4 mm
Column Packing	C8 Reversed Phase (octyldimethyl chain)

Column Temperature	40°C
Mobile Phase	80% acetonitrile 20%(0.1M phosphoric acid solution in water)
Flow Rate	1 ml/min
Detector	fluorescence, excitation at 301 nm emission at 350 nm
Sample Volume	10 µl

It is seen that a very satisfactory separation is obtained; chromatogram (A) was a reference sample containing 0.5 mg/ml of the fungicide thiabendazole; chromatogram (B) was a sample of fruit containing no thiabendazole; chromatogram (C) a sample of fruit that had been spiked with 0.5 mg/ml of the fungicide. This type of sample preparation would be typical for fruit and vegetables, although the specific material that was being analyzed would differ and consequently, the column used and the operating conditions would also differ.

The Preparation of Liquid Samples for LC Analysis

Liquid samples might appear to be easier to prepare for LC analysis than solids, particularly if the compounds of interest are present in high concentration. In some cases this may be true and the first example given below requires virtually no sample preparation whatever. The second example, however, requires more involved treatment and when analyzing protein mixtures, the procedure can become very complex indeed involving extraction, centrifugation and fractional precipitation on reversed phases. In general, however, liquid samples become more difficult to prepare when the substances are present at very low concentrations.

Materials of Interest Present in High Concentration

Even if the components of the sample are at concentrations where direct injection of a suitable volume is possible, all samples should be filtered before placing on the column for the reasons discussed previously. The two examples that will be given are the analysis of lemon juice for carbohydrates and the determination of a fungicide and germicide in a shampoo. The latter is a more complicated procedure as the compound of interest must be hydrolyzed and extracted before it can be separated on a column.

The Determination of Fructose and Glucose in Lemon Juice

The preparation of the lemon juice for analysis was simple, a sample of the lemon juice was filtered and 5 μl placed on the column. The separation was carried out on a novel bonded phase carrying amino-propyl groups. The separation obtained is shown in figure 11.

Figure 11

A Chromatogram of Fructose and Glucose from Lemon Juice

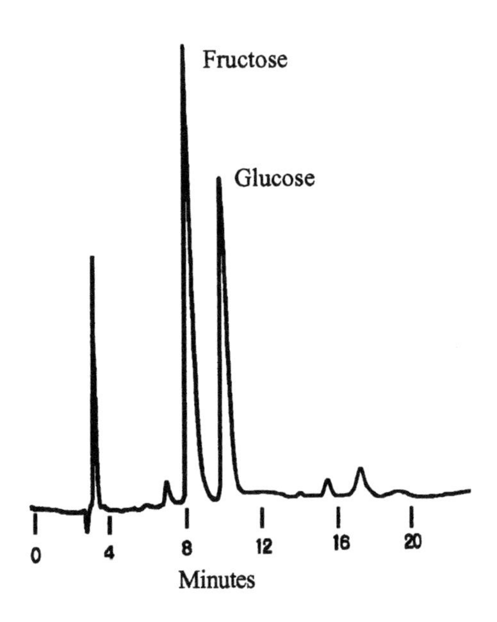

Courtesy of Supelco Inc.

Chromatographic Conditions

Column	L-NH_2
Column Length	25 cm
Column Diameter	4.6 mm
Column Packing	Amino-propyl chains
Mobile Phase	80% acetonitrile 20% water
Flow Rate	1 ml/min
Detector	Refractive index
Sample Volume	5 μl

The dominant interactions that take place with this novel phase are not clear. There will obviously be dispersive interactions with the propyl chain and some polar interactions with the amino group. Whether there are ionic interactions taking place as well is uncertain but, nevertheless, the material affects an excellent separation of glucose and fructose and this remains a baseline separation even when the two sugars are both present at about the 5% level.

The Determination of Suttocide® in Shampoo (10)

This sample preparation involves the decomposition of the fungicide (sodium hydroxymethylglycinate) into sodium glycinate and represents the type of sample that requires significant modification before it can be chromatographed. After decomposition it is then separated by ion exchange chromatography.

An appropriate volume of the shampoo (adjusted to the fungicide content of the sample) was passed through a barium^{2+} cartridge, which had been pre-conditioned with distilled water to remove excess of sulfate ions. The cartridge was washed with distilled water and the sample plus washings were then heated to 60°C for 15 min to convert the sodium hydroxymethylglycinate to sodium glycinate. The sample was then cooled and 25 μl injected onto the column. The chromatogram obtained is shown in figure 12. The peak for sodium glycinate represents 0.18% of the fungicide in the shampoo.

Figure 12

Chromatogram of Sodium Glycinate from Suttocide® Contained in a Shampoo

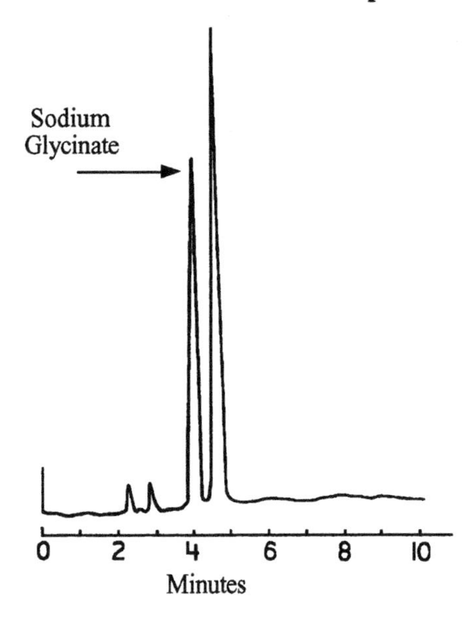

Chromatographic Conditions

Column	Hamilton PRP X100
Column Length	25 cm
Column Diameter	4.6 mm
Column Packing	Proprietary Ion-Exchange
Mobile Phase	0.15M NaOH
Flow Rate	1ml/min
Detector	Integrated amperometric with gold electrode
Sample Volume	25 µl

Detection of the these types of compounds are sometimes difficult as many components of cosmetic products are aliphatic, do not possess a UV chromophore and are not easily reacted to give fluorescent derivatives. Providing the concentration of the component of interest is sufficiently high, then a refractive index detector is often used. If the materials are present at low concentrations then special detectors may be necessary, which, in this case, was the integrated amperometric detector. The separation was achieved by ionic interactions with the ion exchange material but little more can be said as the precise nature of the proprietary ion exchange packing was not reported.

Materials of Interest Present in Low Concentration

The two examples of sample preparation for the analysis of trace material in liquid matrixes are typical of those met in the analytical laboratory. They are dealt with in two quite different ways: one uses the now well established cartridge extraction technique which is the most common; the other uses a unique type of stationary phase which separates simultaneously on two different principles. Firstly, due to its design it can exclude large molecules from the interacting surface; secondly, small molecules that can penetrate to the retentive surface can be separated by dispersive interactions. The two examples given will be the determination of trimethoprim in blood serum and the determination of herbicides in pond water.

The Determination of Trimethoprim in Blood Serum (11)

The analysis demonstrates the elegant use of a very specific type of column packing. As a result, there is no sample preparation, so after the serum has been filtered or centrifuged, which is a precautionary measure to protect the apparatus, 10 μl of serum is injected directly on to the column. The separation obtained is shown in figure 13. The stationary phase, as described by Supelco, was a silica based material with a polymeric surface containing dispersive areas surrounded by a polar network. Small molecules can penetrate the polar network and interact with the dispersive areas and be retained, whereas the larger molecules, such as proteins, cannot reach the interactive surface and are thus rapidly eluted from the column. The chemical nature of the material is not clear, but it can be assumed that the dispersive surface where interaction with the small molecules can take place probably contains hydrocarbon chains like a reversed phase.

Figure 13

The Determination of Trimethoprim in Blood

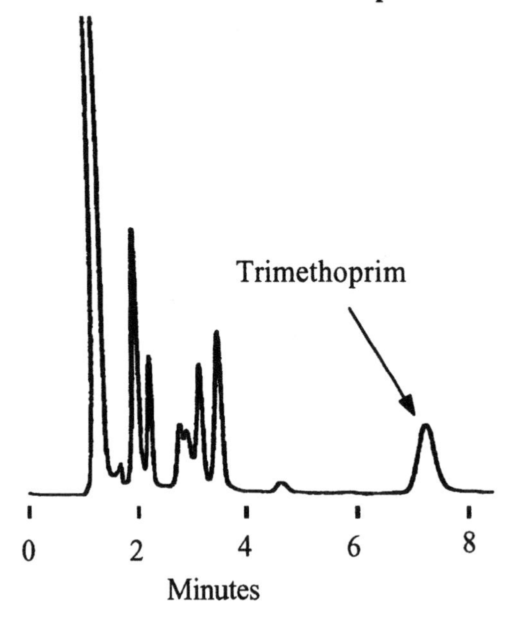

Courtesy of Supelco Inc.

Chromatographic Conditions

Column	Hisep
Column Length	15 cm
Column Diameter	4.6 mm
Column Packing	Proprietary "shielded hydrophobic phase"
Mobile Phase	95% 180mM ammonium acetate and 5%acetonitrile
Flow Rate	1 ml/min
Detector	UV adsorption at 254 nm
Sample Volume	10 µl

The Determination of Triazine Herbicides in Pond Water (12)

This analysis depends on the use of a solid phase extraction cartridge to concentrate the herbicides directly from the pond water and the subsequent separation, by taking advantage of the dispersive interactions that can occur between the herbicides and a reversed-phase. The reverse phase cartridge was first conditioned with 2 ml of methanol, followed by 2 ml of deionized water. Twenty ml of pond water was allowed to percolate through the cartridge, after which the packing was washed with 1 ml of a mixture of 50% water, 37.5% acetonitrile and 12.5% methanol. The sample was eluted from the cartridge with 1 ml of methanol which was collected in a 2 ml volumetric flask and made up to the mark with deionized water. 25 μl of the solution was placed on the column and the chromatogram obtained is shown in figure 14.

Figure 14

Chromatogram of Triazine Herbicides from Pond Water

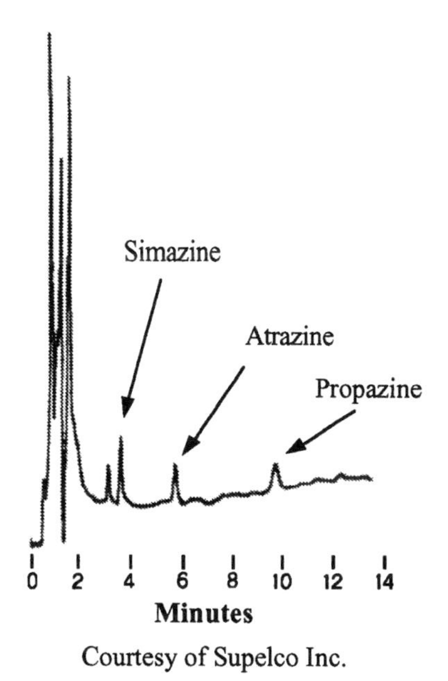

Courtesy of Supelco Inc.

Chromatographic Conditions

Column	LC 8 DB
Column Length	25 cm

Column Diameter	4.6 mm
Column Packing	C8 Reversed Phase (octyldimethyl chain)
Column Temperature	Ambient
Mobile Phase	45% acetonitrile and 55% water
Flow Rate	1.5 ml/min
Detector	UV adsorption at 254 nm
Sample Volume	25 μl

The solutes were separated by standard reverse phase processes where differential interactions occur between the dispersive centers of the solute molecules and the hydrocarbon chains of the reversed phase. The concentration of the herbicides in the pond water was 10 ppb.

The Preparation of Liquid/Solid Samples for LC Analysis

Samples that contain two phases present a special problem depending on the site of the materials of interest. If the substances are known to be associated with one phase only, the sample procedure is simply to separate the two phases by filtration or centrifugation and treat the sample as a liquid or solid sample depending on the phase that contains the materials to be determined. However, if the materials of interest are distributed between the two phases, some in solution and some adsorbed on the surface of the solid, then special extraction procedures will be necessary.

The whole sample can be extracted where this is possible or, again, the two phases separated and both extracted by different techniques. As stated before, adsorption of the substances to be determined on the solid phase can become difficult when they are present at very low concentrations and the adsorbed material is a significant proportion of the total sample.

Typical areas where two phase samples commonly occur are foodstuffs, biological tissue and sludges from environmental tests.

Materials of Interest Present in High Concentration

Two-phased samples, where the components of interest are present in high concentration in the liquid, can often be dealt with by simple filtration as the amount of material adsorbed on the surface of the solid phase, relative to that in the liquid phase is likely to be insignificant. Alternatively if the material is dispersed as a solid throughout the solid phase, then the sample can be filtered and the solid extracted exclusively.

The two examples given here will be the determination of flavones in grapefruit juice and the measurement of histamine in rat brain.

The Determination of Flavones in Grapefruit Juice (13)

The sample preparation was very simple: the sample was centrifuged to remove plant cell residues and 10 µl of the clear juice was placed on the column. This type of separation is common with fruits, vegetables and juices and samples can be obtained by preliminary homogenizing the total tissues and then centrifuging. If it is suspected that the residue still contains significant quantities of the substances of interest, then it can be washed with water or if necessary with solvents and the washings combined with the separated supernatant liquor. The results obtained are shown in figure 15.

Chromatographic Conditions

Column	LC 18
Column Length	15 cm
Column Diameter	4.6 mm
Column Packing	C18 Reversed Phase (octyldimethyl chain) Particle size 3 µm
Mobile Phase	20% acetonitrile 80% 0.5 acetic acid in water
Flow Rate	1ml/min
Detector	UV adsorption at 280 nm
Sample Volume	10 µl

Figure 15

Chromatogram of the Flavones in Grapefruit Juice

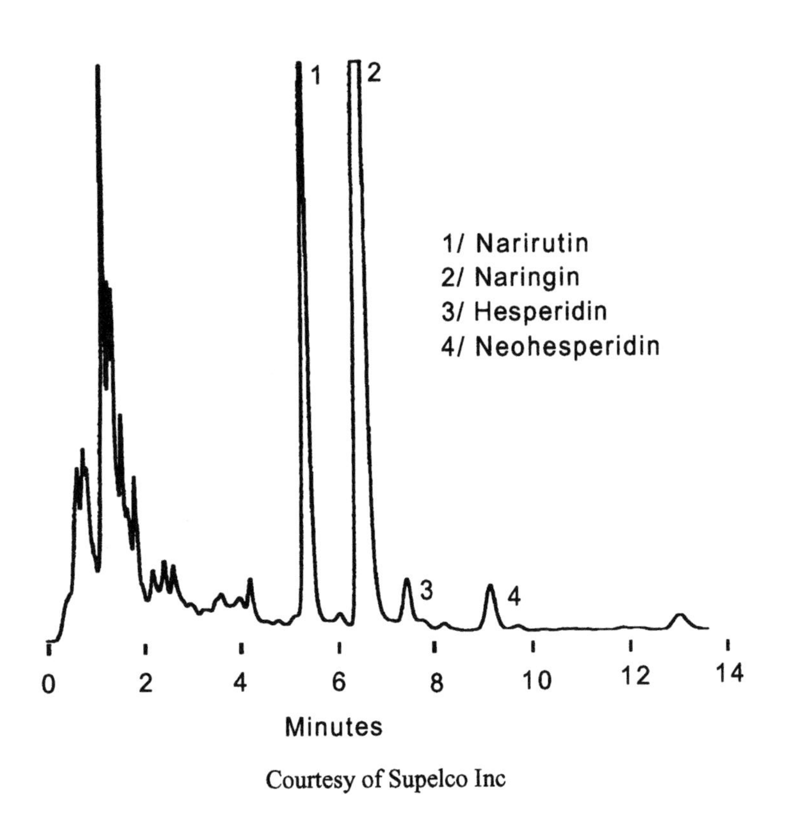

Courtesy of Supelco Inc

The Determination of Histamine in Rat Brain(14)

A weighed quantity of rat brain was treated with an appropriate volume of 0.4 M perchloric acid and homogenized. The homogenate was then centrifuged for 20 min at 20,000 rpm and 1 ml of the supernatent liquid removed.

To the 1 ml sample 100 µl of concentrated potassium hydroxide was added and the mixture centrifuged again for 2 min. The supernatent liquid was heated in boiling water for 4 min to precipitate the protein, cooled and again centrifuged.

Figure 16

Chromatogram Showing Histamines from Rat Brain Tissue

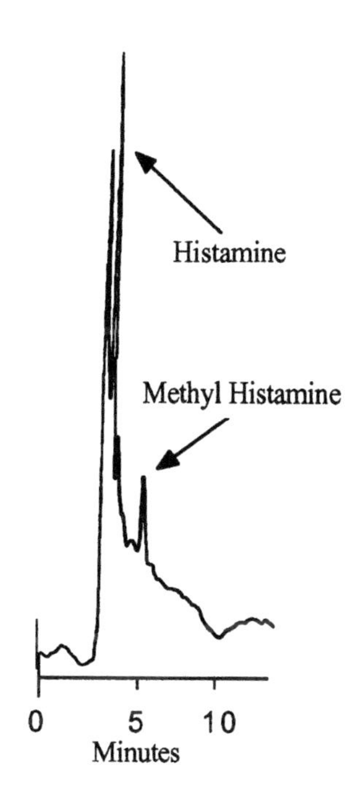

Chromatographic Conditions

Column	C18
Column Length	25 cm
Column Diameter	4.6 mm
Column Temperature	ambient
Column Packing	C18 Reversed Phase (octadecyldimethyl chain)
Mobile Phase	79% (0.12 M NaH_2PO_4, 0.1 M NaOH, 19 μM of lauryl sulfate and 5%methanol in water) and 21% water

A sample of the supernatent liquid was diluted with phosphate buffer and an appropriate volume placed on the column. The chromatogram obtained is shown in figure 16.

Complete resolution was not achieved due to the carryover of interfering substances which frequently occurs when separating the components of biological samples. The column carried a reverse phase, but as the mobile phase contained low concentrations of lauryl sulfate, some would have adsorbed on the surface of the stationary phase and significantly modified its interacting properties. The retention mechanism is likely to have involved both ionic interactions with the adsorbed ion exchanger together with dispersive interactions with any exposed areas of the reverse phase.

Materials Present in Samples at Low Concentration

Samples usually require both an extraction and a concentration technique, when the solutes of interest are present in the mixture at low concentrations. Sometimes the two stages are combined in one, as in the use of solid phase extraction cartridges. In other cases a two stage procedure is necessary, firstly the materials are extracted and secondly they are concentrated. The examples given below include both types of procedure. The sample preparation for the analysis of metanephrine (15) in urine is a very complex procedure, involving both solvent extraction of interfering materials and concentration by an ion exchange column. The determination of phenol in water (16) is more traditional and uses a short cartridge type column to effect the extraction and concentration.

The Determination of Metanephrine and Normetanephrine in Urine

Twenty ml of urine was adjusted to a pH of 1.0 in a glass tube and placed in boiling water bath for 30 min. The urine was then cooled and 5 ml of a 60% toluene/40% amyl alcohol added. After shaking for 2-3 minutes, the mixture was centrifuged briefly and the upper solvent layer, containing the urinary pigments, separated and discarded. The remaining aqueous layer was diluted with 20 ml of boric acid solution (0.65 mol/liter) and the pH adjusted to 6.5 with 5N NaOH. The solution was passed through a column filled with Amberlite CG 50.

eluted with 5 ml of ammonia solution (4 mol/liter) into a glass tube containing 2.5 g of sodium chloride. The solution was then extracted twice with 4 ml of a 2:1 mixture of ethyl acetate and acetone and shaken for 5 minutes.

Figure 17

Estimation of Metanephrine and Normetanephrine in Urine

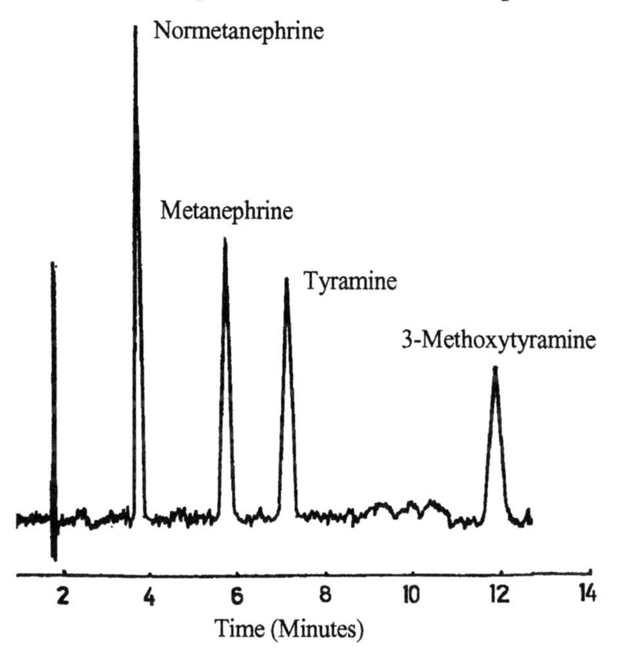

Chromatographic Conditions

Column	Seperon Six (particle diameter 5 µm)
Column Length	15 cm
Column Diameter	3.2 mm
Column Packing	C18 Reversed Phase (octadecyldimethyl chain)
Mobile Phase	5% methanol, 95% aqueous solution of 0.1 m/l NaH_2PO_4 and 1 mM/l of sodium ethylene-diamine tetra-acetate
Flow Rate	1.6 ml/min
Detector	UV adsorption at 280 nm
Sample Volume	10 µl

The mixture was then centrifuged and a 4 ml aliquot of the organic layer evaporated on a water bath at 40°C under a stream of nitrogen. The residue was dissolved in 50 μl of mobile phase and 5-10 μl injected onto the column. The separation obtained is shown in figure 17 and it is seen that an excellent resolution was obtained. As the mobile phase contained ethylenediaminetetra-acetate some would almost certainly be adsorbed on to the surface of the reversed phase because the methanol content was only 5%. As a consequence, the separation was probably achieved by an ion exchange mechanism with the adsorbed ion exchanger. However, in addition there were almost certainly some dispersive interactions occurring between the solutes and the exposed reversed phase not covered with the ethylenediamine-tetra-acetate. This example also demonstrates how complex a sample preparation procedure can be. In general, in the development of the analytical method great effort is made to make the procedure as simple as possible but, unfortunately, this is not always feasible.

The Determination of Phenol in River and Waste Water

Phenol is a common water pollutant as it is a by-product of many commercial processes. Moreover, it is also very toxic to all forms of plant life. Lehotay, Baloghova and Hatrik developed a procedure where the material was extracted and concentrated by a pre-column which was, in effect, acting as a solid phase extraction cartridge. The system was automated in a manner similar to that already described. The pre-column was washed with distilled water and 1 ml of the sample pumped through it. The pre-column was then washed with more distilled water and connected in line with the LC column. The separation was developed using a solvent gradient starting with 40% methanol in water and ending with pure methanol.

The results obtained are shown in figure 18, the upper chromatogram representing the polluted river water and the lower chromatogram a sample taken after clean-up treatment. The level of phenol present in the lower chromatogram was about 34 ppt. The separation was monitored by an electrochemical detector which had a very high specific response to phenol. The column used was packed with C18 reverse phase and so the separation was again based on dispersive interactions between the phenol and the hydrocarbon chains of the bonded phase.

Figure 18

The Determination of Phenol in Waste Water

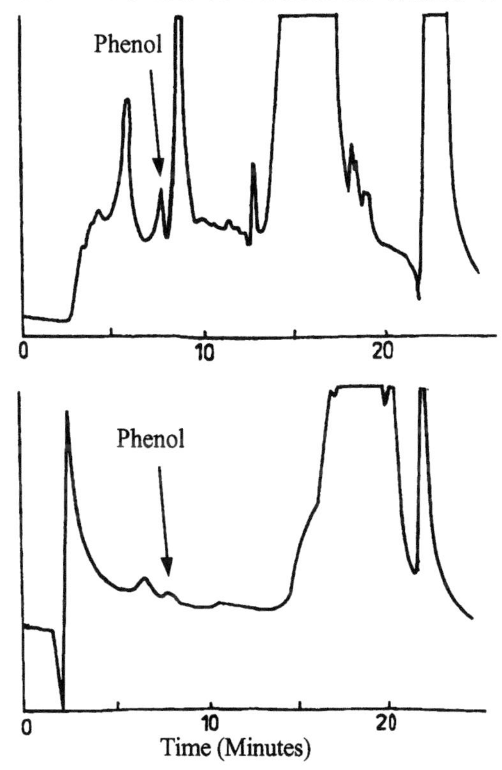

General Comments on Choice of Mobile Phase

Excluding those examples where adsorbed ion exchange reagents have been used, it is seen from the examples given that most of the separations have used reversed phases as the stationary phase. Although, as a consequence, the retentive interactions of the solutes with the stationary phase have been largely dispersive, selectivity has been adjusted by changing the nature and strength of the interactions in the mobile phase.

Increasing the solvent concentration increases the level of the competitive dispersive interactions that take place in the mobile phase. Use of a more dispersive solvent, such as tetrahydrofuran, will also achieve the same effect. In the same way, use of a more *polar* solvent such as methanol will increase the competitive *polar* interactions in the mobile phase.

Thus, by careful choice of solvents, evoked by an understanding of the essential role played by the different types of molecular interactions in the chromatographic process, the solutes of interest cannot only be separated, but also eluted in a reasonable time.

As already stated, the newcomer to LC can obtain considerable help from the many publications available, but as experience is gained the need to resort to the literature will become less and less. The initial tendency to change the mobile phase composition and "see what happens" should be avoided; any solvent change should be the result of rational deliberation based on sound physical chemical reasoning. There are a number of computer programs available that assist in optimizing the solvent composition for a specific separation. Usually the basic solvent system must be chosen first and then a number of solvent changes are made (sometimes suggested by the computer program). The results obtained by the different solvents are entered into the computer and the optimum solvent mixture to effect the separation is suggested. Such programs may well help the analyst with little prior knowledge but, with experience, and a rational approach, the analyst is likely be able to arrive at the required optimum solvent mixture without the aid of the computer.

The use of other important phase systems such as exclusion media, ion exchange media and polar stationary phases such as silica gel have not been discussed as this chapter is primarily concerned with sample preparation. The last chapter will give examples of the use of these other phase systems and explain the separations obtained on a basis of molecular interactions and, at that time, the subject of solvent choice will again be discussed.

Derivatization Techniques

In analytical LC there are two primary reasons why chemical derivatization of the sample constituents would be necessary, and they are 1) to enhance the separation and 2) to increase the sensitivity of detection. Under certain circumstances, derivatization can also be used to reduce peak asymmetry, i.e. to reduce tailing, or to improve the stability of labile components so that they do not re-arrange or decompose during the chromatographic process. However, sensitivity enhancement is the most common goal of derivatization. For example, aliphatic alcohols that contain no UV chromaphore can be reacted with benzoyl chloride to form a benzoic ester.

$$C_6H_5\text{-CO-Cl} + \text{R-O-H} = C_6H_5\text{-CO-O-R} + \text{HCl}$$

The aromatic nucleus adsorbs in the UV and thus, the derivative can be detected by a UV detector. This is the most common type of chemical derivatization but the derivative may be chosen to be appropriate for different types of detector. For example, the solute can be reacted with a fluorescing reagent, producing a fluorescent derivative and thus be detectable by the fluorescence detector. Alternatively, a derivative can be made that is easily oxidized and, consequently, would be detectable by an electrochemical detector.

In any derivatization procedure certain criteria must be met. Firstly, the reaction must be rapid and quantitative, or at the very least reproducible. Secondly, the reagents must be easily separated from the final products and they must produce a *single* product from each component. Thirdly, the derivatives must have detection properties that differ from those of the original components and the capability for their subsequent chromatographic separation must be preserved.

Chemical derivatization can be carried out before the separation (pre-column derivatization) or after the separation and before detection (post-column derivatization). If derivatization is carried out prior to separation, then a phase system must now be selected to separate the *derivatives* which is likely to differ considerably from the phase system used to separate the original solutes.

Pre-Column Derivatization

Fluorescence Reagents

An excellent discussion on derivatization techniques has been given by Lawrence (17) including a detailed discussion on pre-column derivatization (18) and post-column derivatization (19). Probably, the more popular procedures are those that produce fluorescing derivatives to improve detector sensitivity. One of the more commonly used reagents is dansyl chloride (20), 5-dimethylamino-naphthalene-1-sulphonyl chloride (sometimes called DNS-chloride or DNS-Cl). The reagent reacts with phenols and primary and secondary amines under slightly basic conditions forming sulphonate esters or sulphonamides.

SO_2— Cl   + R-O-H = SO_2— O-R   + HCl

N   R   R   N   R   R

or

SO_2— Cl   + R-NH_2 = SO_2—O-NH-R   + HCl

N   R   R   N   R   R

wherein the dansyl reagents (R) is a methyl group.

It should be noted that there are other dansyl reagents available that have somewhat different attributes. The reagents are designated as follows,

R = C_2H_5	ethansyl chloride
R = n-C_3H_7	propansyl chloride
R = n-C_4H_9	bansyl chloride
R = n-C_5H_{11}	pentansyl chloride

The quantum efficiency of the dansyl derivatives is high (that is the proportion of incident light falling on the derivative that results in the emission of light of a longer wavelength is relatively large); the reagent itself does not fluoresce. However, the hydrolysis product of the reagent, dansylic acid has strong fluorescence and can cause interference with water soluble derivatives that require an aqueous mobile phase to achieve separation. In some cases the interfering materials are separated from the derivatives of interest by the chromatographic process. Detection limits are around 10^{-9} g and the optimum wavelength of the excitation light is 350-370 nm and that of the emitted light between 490 and 540 nm. Examples of the use of the dansyl reagents have been reported for amino acids, (21), alkaloids (22), barbiturates (23) and pesticides (24).

NBD-chloride (4-chloro-7-nitrobenz-2,1,3-oxadiazole) reacts with aliphatic primary and secondary amines to form derivatives with strong fluorescence. Fortunately phenols, thiols and aromatic amines give derivatives that either do not fluoresce at all or give very weak fluorescence. As a consequence NBD-chloride is specific for aliphatic primary and secondary amines. The reaction is carried out under basic conditions and the products can be extracted into polarizable solvents such as benzene or polar solvents such as ethyl acetate. The fluorescence from the derivatives can be quenched with water and it follows that mobile phases containing water would not be suitable for their separation. The sensitivity of the detector to such derivatives is $<10^{-9}$ g. The wavelength of the excitation light is about 480 nm and that of the emitted light is about 530 nm. Examples have been given of its use in the analysis of amino acids (25), amphetamines (26), alkaloids (27) and nitrosamines (28).

Another reagent that readily forms fluorescent derivatives with primary amines is o-phthalaldehyde (trade name "Fluoropa"). The reaction proceeds in aqueous solution in the presence of a mercaptan at a pH of 9-11 producing an isoindole.

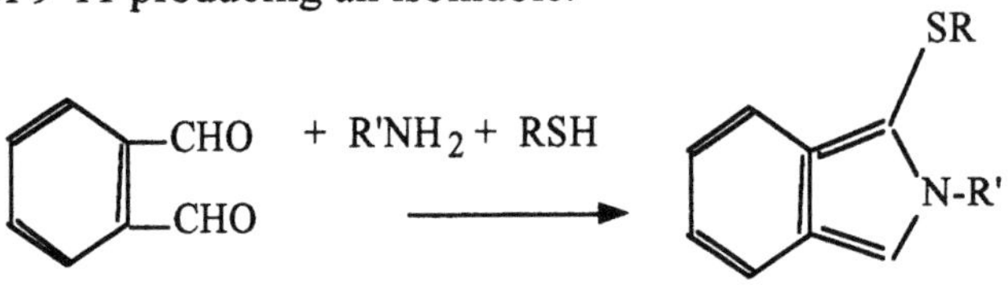

$$\text{C}_6\text{H}_4(\text{CHO})_2 + \text{R'NH}_2 + \text{RSH} \longrightarrow \text{isoindole (SR, N-R')}$$

The derivatives have an optimum fluorescence at an excitation wavelength of 340 nm and an emission wavelength of 455 nm. The adduct is relatively stable at a pH of 9-11 but it rapidly degrades to a non-fluorescent residue at low pH values. Consequently, when used as a pre-column derivatizing reagent the pH of the mobile phase should be kept fairly high. o-phthalaldehyde has been employed for derivatization in the analysis of dopamine (29), catecholamines (30) and histamines (31).

Another commercially available fluorescing reagent, "Fluorescamine", (4-phenylspiro(furan-2-(3H),1'-phthalan)3,3'-dione) reacts directly with a primary amine in aqueous acetone at a pH 8-9.

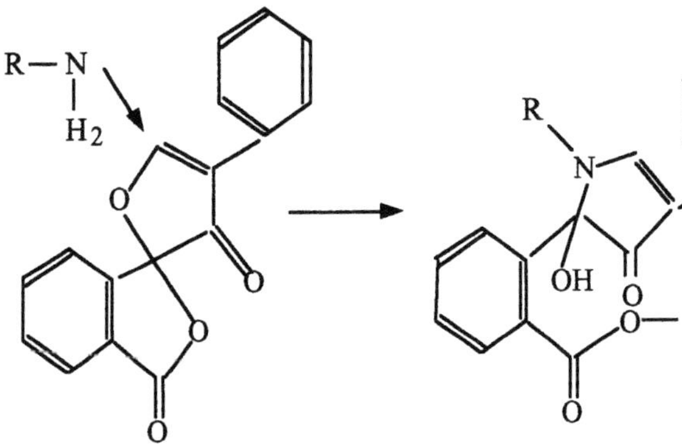

Excess of the reagent hydrolyses to a non-fluorescent residue and the reagent itself does not fluoresce. The optimum wavelength of the excitation light is 390 nm and that of the emitted light 475 nm. This regent is, however, less sensitive than Fluoropa and the derivative is unstable consequently, it must be injected onto the column immediately after formation if used in pre-column derivatization. It has been used successfully in the separation and analysis of polyamines (32), catecholamines (33) and amino acids (34).

UV Adsorption Enhancers

The use of UV adsorption enhancers as reagents that introduce a UV chromaphore into a molecule that is transparent in the UV wavelength range has already been briefly discussed. The two most common reagents are the phenyl and methyl isothiocyanates. These reagents react with amino acids to form thiodantoins.

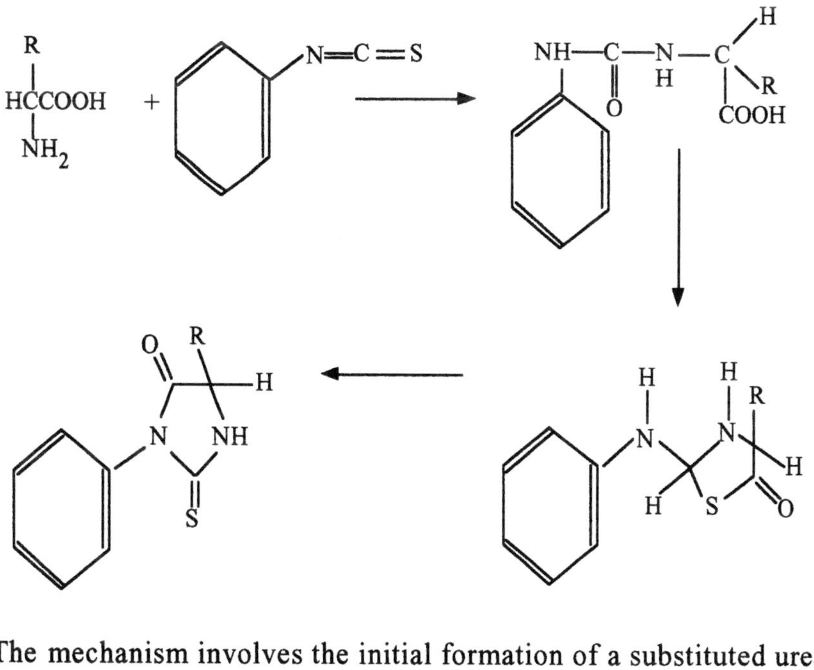

The mechanism involves the initial formation of a substituted urea followed by ring closure to form the thiohydantoin. The amino acid is dissolved in 60% aqueous pyridine containing the phenylisothiocyanate and the mixture heated to 40°C for one hour to complete the reaction. The derivatives are stable under acid conditions and can, therefore, be used for amino acid sequencing (35,36) in peptide hydrolysis. The phenyl derivatives can be detected at much lower concentrations than the methyl analogs. At 260 nm a UV detector can distinguish concentrations as low as 10^{-9} g/ml from background noise.

The use of benzoyl chloride has already been mentioned but, p-methoxy-benzoylchloride and 3,5-dinitro-benzoyl chloride have much greater extinction coefficients owing to their ring substitution than p-nitrobenzoyl chloride, and thus, give much greater detector sensitivity. These reagents have been employed in the analysis of hydroxysteroids (37), digitalis glycosides (38), carbohydrates (39) and amphetamines (40).

Another reagent, 2,4-dinitrofluorobenzene, also known as Sanders Reagent, has been used in a similar manner in the analysis of amino acids and amines (41).

R
HCCOOH +
NH_2
F
NO_2
NO_2
R
HCCOOH
NH
NO_2
NO_2
+ HF

The reagent reacts with primary and secondary amines and phenols but not aliphatic alcohols and is, thus, particularly useful for determining phenols in the presence of aliphatic hydroxy compounds.

Reagents that provide UV adsorptive derivatives of carboxylic acids are fairly numerous. The preparation of the simple benzyl esters by reacting the carboxylic ion with alkyl halides or diazo compounds has been unsuccessful due to their having unacceptable toxicity. The triazenes such as 1-benzyl-3-p-tolyltriazene react with free fatty acids to give the corresponding benzyl esters (42).

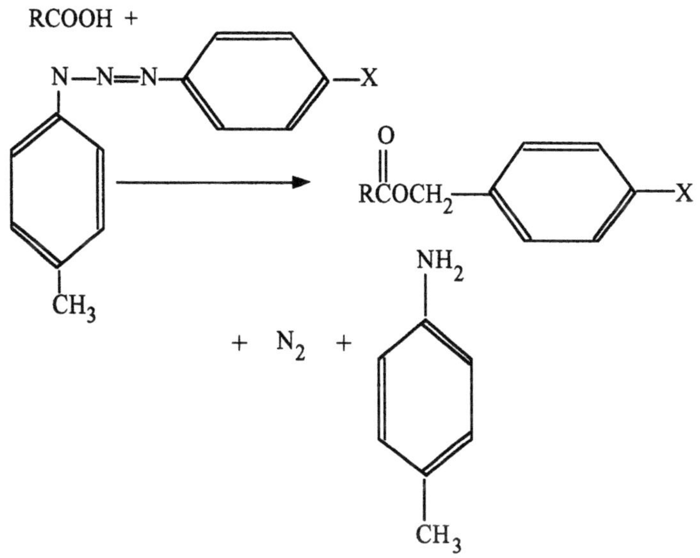

RCOOH +

N—N=N—⟨ring⟩—X

CH_3

⟶ $RCOCH_2$ (C=O) —⟨ring⟩—X

+ N_2 + NH_2 —⟨ring⟩— CH_3

In the case of 1-benzyl-3-p-tolyltriazene, X=H.

The normal procedure is to dissolve the acid in ether, the triazine is added and the mixture maintained at 35°C for three hours. The mixture is then washed with successive amounts of hydrochloric acid to remove the p-toluidine and excess reagent and the final ether solution can then be injected onto the LC column. The derivatization procedure reduces the difference in potential phase interactions between the different carboxylic acids and thus, the separation of the acid derivatives may be more difficult than the original acids. There are some disadvantages to the triazines as derivatizing agents: they are carcinogenic, the evolution of nitrogen must not be allowed to be too vigorous and causes sample loss, they are expensive and under some circumstances the formation of by-products can confuse the analysis.

The most popular reagent for the formation of aldehyde and ketone derivatives is 2,4-dinitrophenylhydrazine which forms hydrazones containing strong chromophores.

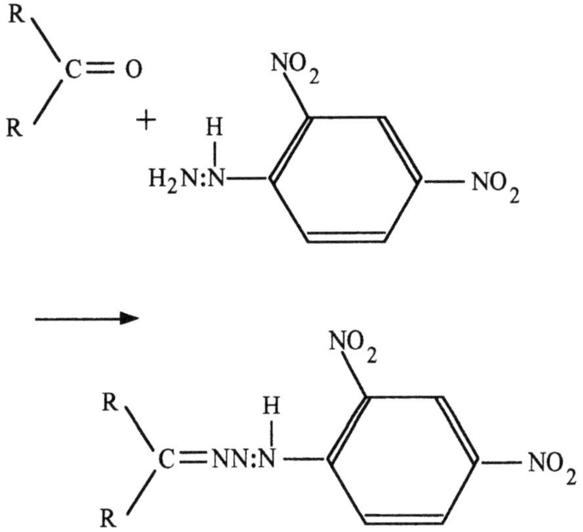

R, R C=O + $H_2N{:}N(H)$— [2,4-$(NO_2)_2C_6H_3$] ⟶ R, R C=NN:N(H)— [2,4-$(NO_2)_2C_6H_3$]

The reaction is carried out by dissolving the aldehyde or ketone in methanol, acidifying the solution with concentrated hydrochloric acid and then adding a solution of the reagent in methanol. The reaction is allowed to proceed for about 2 hours at 50-60°C.

2,4-Dinitrophenyl-hydrazine has been successfully employed in the analysis of simple aldehydes, substituted aldehydes, glyoxal and gluteraldehyde (43-45), all the isomers of the C3 to C7 aliphatic ketones (44,45) and in the determination of formaldehyde in tobacco smoke (46).

There is a wide range of reagents available for derivatization and the analyst is again referred to the books by Frei and Lawrence (17) and Karl Blau and John Halket (47) for further reading. The references given here have been chosen as those that are most likely to include the complete details of the derivatizing procedures. They were not chosen as the most contemporary examples of analyses employing derivatization techniques.

Post-Column Derivatization

In post-column derivatization the chromatographic system is modified to allow the reagent to mix with the column eluent, give the reaction mixture sufficient time to complete and finally pass the reaction mixture to the detector.

Figure 19

A Chromatograph System Including a Post-Column Reactor

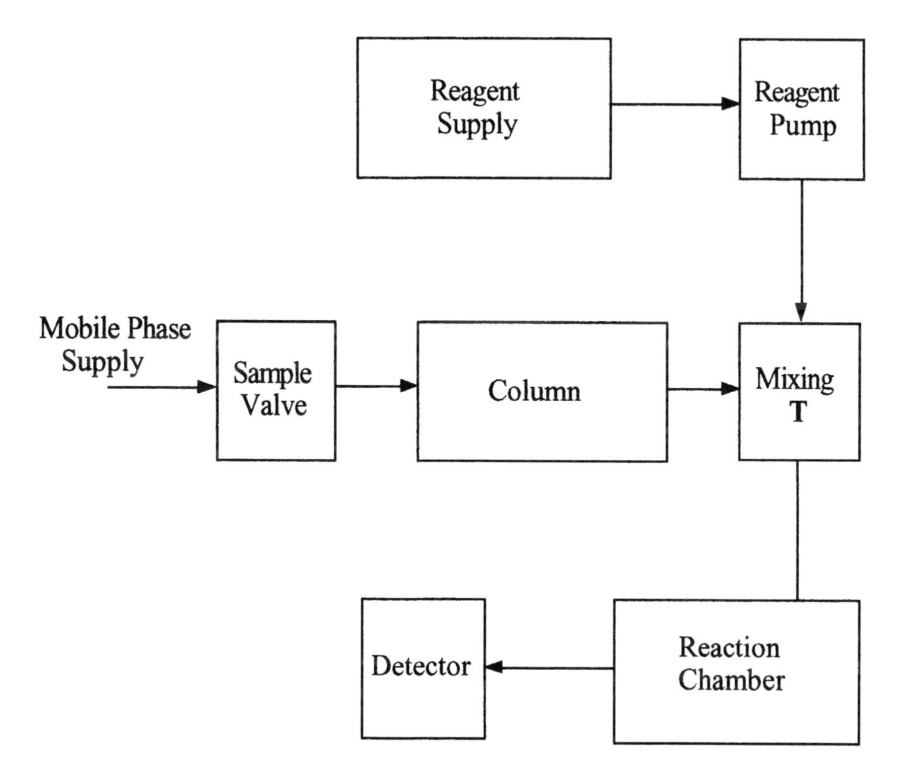

A separate reagent pump is required for the reagent supply, some form of blending **T** that combines the reagent and eluent streams and ensures an adequate reagent-eluent mixing and finally a delay coil or tube. A block diagram of the lay-out of a post-column reactor is shown in figure 19. The essential part of the post-column reaction system is the mixing **T** as this device must be designed in such a way that there is no segregation of the reagent and the column eluent and both are intimately mixed. If there are any significant pulses in either the reagent flow or the column flow, alternate segments of reagent and column eluent will enter the reactor and, incomplete mixing of this kind, will significantly slow the reaction. It is also important that the eluent conduit system between the column and the detector, including the mixing **T** and the reactor, causes minimum peak dispersion as this would degrade the separation. Invariably, some dispersion does occur in the post-column reactor, but by careful design, the dispersion can be contained to an acceptable level.

One of the factors that is extremely important in the reactor design is the shape of the mixing **T**. Simple right-angled **T** joints are very inefficient and according to Zech and Voelter (48) the mixing junction should take the form of that shown in figure 20.

Figure 20

Efficient Form of a Mixing T

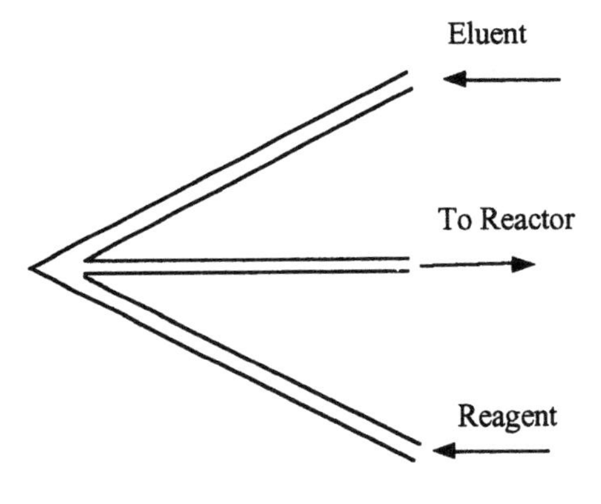

The angle between the eluent and the reagent tube to the reactor tube was 30°. Efficient mixing resulted from the eddies that formed at the point of mixing due to the reversal of flow direction by the eluent and reagent streams.

The reactor can consist of a short packed tube or a length of coiled tube. Open tubes can give very serious band dispersion as already discussed and, if a tube is used for the reactor, it should be constructed of low-dispersion tubing. Low dispersion tubing will not only reduce band dispersion but will also produce highly efficient mixing and thus accelerate the reaction.

To minimize pulses in the reagent flow, for the reasons already mentioned, a syringe pump is normally used for the reagent supply. Alternatively, if a piston pump is used, then an efficient pulse dampener should be inserted between the pump and the mixing **T**.

Post-column reaction is a common feature of many special types of analyses, the most well-known being the amino acid analyzer that uses ninhydrin with a post-column reactor to detect the separated amino acids. In general, derivatization and post-column reactor systems are techniques of last resort. In some applications they are unavoidable, but if possible, every effort should made to find a suitable detector for the actual sample materials before resorting to derivatization procedures.

References

1/ Application Sheet No 909100-0181 Supelco Inc.

2/ Application Sheet No 909100-0271 Supelco Inc.

3/ R.P.W.Scott and P.Kucera,*J.Chromatogr.***185**(1979)27.

4/ R.P.W.Scott and P.Kucera,*J.Chromatogr.***169**(1979)51.

5/ Application Sheet No 909100-0057 Supelco Inc.

6/ Application Sheet No 103 Whatman Inc.

7/ Application Sheet No 909100-0019 Supelco Inc.

8/ Noboru Motohashi,Hideo Nagashima and Oger Meyer, *J.Liquid Chrom.*,**14(19)**(1991)3591.

9/ Application Sheet No 311001-0005 Supelco Inc.

10/ R.M.Ianniello, *J.Liquid Chrom.*,**15(17)**(1991)3045.

11/ Application Sheet No 309003-0009 Supelco Inc.

12/ Application Sheet No 909200-0048 Supelco Inc.

13/ Application Sheet No 911100-0007 Supelco Inc.

14/ B.Washington, Kinh Nguyen and R.F. Ochillo, *J.Liquid Chrom.*,**16(5)**(1993)1195.

15/ P.Kaplan, M. Tejral and I.Nentwich, *J.Liquid Chrom.*, **15(14)**(1992)2561.

16/ J.Lehotay,M.Balogova and S.Hatrik, *J.Liquid Chrom.*, **16(5)**(1993)999.

17/ "Chemical Derivatization in Analytical Chemistry" (Eds R.W.Frei and J.F.Lawrence), Plenum Press,New York (1981).

18/ L.A.Sternson, in "Chemical Derivatization in Analytical Chemistry" (Eds R.W.Frei and J.F.Lawrence), Plenum Press, New York (1981)127.

19/ R.W.Frei, in "Chemical Derivatization in Analytical Chemistry" (Eds R.W.Frei and J.F.Lawrence), Plenum Press, New York (1981)211.

20/ B.S.Hartly and V.Maney,*Biochem.Biophys. Acta.*,**21**(1956)21.

21/ T.Kinoshita,F.Linuma and K.Atsumi, *Chem.Pharm.Bull.*, **23**(1975)1166.

22/ R.W.Frei,W.Santi and M.Thomas,*J.Chromatogr.*,**116**(1976)365.

23/W.Dunges,G.Naundorf and N.Seiler,*J.ChromatogrSci.*,**12**(1974)655.

24/ J.F.Lawrence, and R.W.Frei, *J.Chromatogr.*,**66**(1972)93.

25/ P.B.Gosh and M.W.Whitehouse, *Biochem. J.,* **108**(1968)155.

26/ J.Monforte,R.J.Bath and I.Sunshine.,*Clin. Chem.*,**18**(1972)329.

27/ J.Reisch,H.J.Kommert and D.Clasing,*Pharm.Zig.*,**115**(1970)754.

28/ H.J.Klimisch and D.K.Anmbrosious,*J.Chromatogr.*,**121**(1976)93.

29/T.P.Davis,G.W.Gehrke,T.D.Cunningham,K.C. Kuo,K.O.Gerhadt, H.D.Johnson and C.H.Williams,*Clin.Chem.*,**24**(1978)261.

30/ L.D.Mell,A.R.Daslar and A.B.Gutafrom,*J.Liq.Chromatogr.* **1**(1978)261.

31/ R.E.Subden,R.G.Brown and A.C.Noble,*J.Chromatogr.* 166(1978)310.

32/ K. Samejima. *J.Chromatogr.* **96**(1974)250.

33/ G.Schwedt, *J. Chromatogr.* 118(1976)429.

34/ W.McHigh,R.A.Sandman,W.G.Haney,S.P.Sood and D.P.Wittmer, *J.Chromatogr.*, **124**(1976)376.

35/ V.M.Spepanov,S.P.Katrukha,L.A.Baratova, L.P.Belyanova and V.P.Korzhenko, *Anal Biochem.,* **43**(1971)209.

36/ J.Elion,M.Downing and K.Mann,*J.Chromatogr.*,**155**(1978)436.

37/ P.J.Porcaro and P.Shubiak, *Anal. Chem.* **44** (1972) 1865.

38/ F.Nachtmann,*Z. Anal.Chem.*,**282** (1976)209.

39/ C.R.Clark,J.D.Teague,M.M.Wells and J.H.Ellis, *Anal. Chem*. **49**(1977)912.

40/ R.Schwyzer and H.Kappler,*Helv.Chim.Acta.*,**46**(1974)24.

41/ H.Beyer and U. Schenk, *J. Chromatogr*.,**39**(1969)482.

42/L.R.Politzer,G.W.Griffin,B.J.Dowty and J.L.Laseter *Anal.Lett.*6(1973)539.

43/ L.J.Papal and L.P.Turner,*J.Chromatogr.Sci.*,**10**(1972)747.

44/ S.Semlin, *J.Chromatogr.*,**136**(1977)271.

45/ R.W.Frei and J.FLawrence,*J.Chromatogr.*,**83**(1973)321.

46/ C.T.Mansfield,B.T.Hodge,R.B.Hedge and W.C.Hamlin, *J.Chromatogr. Sci.*,**15**(1977)301.

47/ "*Handbook of Derivatives for Chromatography*", (Ed. Karl Blau and John Halket) John Wiley and Sons, New York-Chichester-Brisbane-Toronto-Singapore (1993).

48/ K.Zech and W.Voelter, *Chromatographia*, **8**(1979)350.

8
Qualitative and Quantitative Analysis

The object of an LC analysis is to establish the probable identity, and determine the precise amount, of each of the pertinent components present in the sample. The pertinent components may include all the substances present in the mixture or only those of specific interest. The identity of the peak is determined from its position on the chromatogram, that is, the time required for it to be eluted, whereas, the quantity of a component present is determined from the peak height or peak area. It must be emphasized that a single LC analysis on a *hitherto unknown sample* can not unambiguously confirm the presence of a particular compound on the basis of retention data alone. Retention data, whether it is corrected retention volume, capacity factor or the separation ratio of the solute to that of a standard, can only indicate the probability of substance identity. Retention data from a second analysis, using a different phase system, increases the confidence level but absolute verification requires confirmation by another analytical technique. This might include infrared spectrometry, mass spectrometry or nuclear magnetic resonance spectroscopy. Such evidence would be essential for litigation purposes.

There are in-line LC/spectroscopic systems available, but in most cases it is easier to carry out a semi-preparative separation, collect the material and carry out the spectroscopic examination off-line. However, for routine quality control analyses, where the *sample characteristics are already well established*, retention data can be quite sufficient for identification purpose.

Qualitative Analysis

In LC both quantitative and qualitative accuracy depends heavily on the components of the sample being adequately resolved from one another. The subject of resolution has already been discussed, but it is necessary to consider those areas where uncertainty can still arise. Unfortunately, unless the analyst is aware of the pitfalls and how to deal with them, false assumptions of resolution can be made very easily.

Consider the liquid chromatography peak shown in figure 1.

Figure 1 **A Single LC Peak**

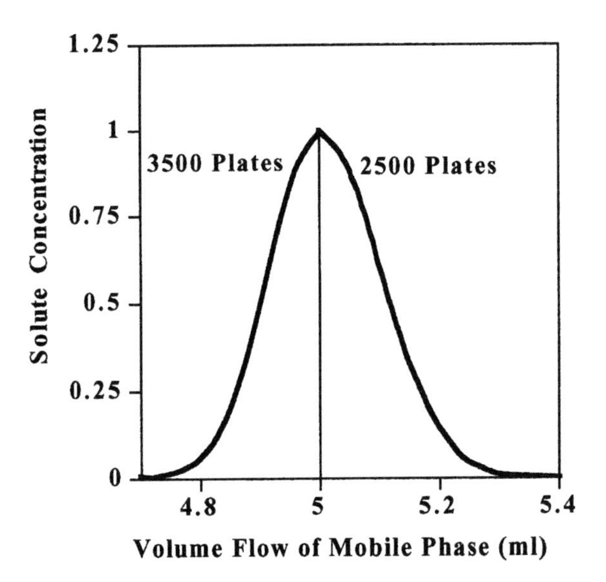

It is seen that the peak shown in figure 1 is asymmetric, which is typical of many peaks eluted from LC columns. This particular peak is not grossly asymmetric, the front half of the peak having an efficiency equivalent to 3500 theoretical plates and the latter half an efficiency of

2500 theoretical plates. This situation frequently occurs in LC and can be caused by a number of different effects. The two major causes of peak asymmetry have already been touched upon and can arise from a difference in peak dispersion occurring in the front half of the peak relative to the rear half or to the solute distribution coefficient being different for the two halves of the peak.

Re-iterating the HETP equation (10) given on page 104,

$$H = 2\lambda d_p + \frac{2\gamma D_m}{u} + \frac{f_1(k')d_p^2}{D_m}u + \frac{f_2(k')d_f^2}{D_S}u \qquad (1)$$

where the symbols have the meanings previously defined.

Now, it was also shown on page 145 that for a given column, solute, mobile phase and flow rate, equation (1) can be reduced to an alternative abbreviated form which is given as follows,

$$H = a + bD_m + \frac{c_1 d_p^2}{D_m} + \frac{c_2 d_f^2}{D_S} \qquad (2)$$

where $a = 2\lambda d_p$, $b = \frac{2\gamma}{u}$, $c_1 = f_1(k')u$ and $c_2 = f_2(k')u$.

On page 6, it was shown that in the front half of the peak, there will be a net transfer of solute from the mobile phase to the stationary phase and thus the resistance to mass transfer in the mobile phase will dominate. At the rear half of the peak there is a net transfer of solute from the stationary phase to the mobile phase and in this case the resistance to mass transfer in the stationary phase will dominate. Then if the resistance to mass transfer in the stationary phase is greater than that for the mobile phase, the rear part of the peak will be broader than the front half. In which case,

$$\frac{c_1 d_p^2}{D_m} < \frac{c_2 d_f^2}{D_S} \qquad (3)$$

Consequently, the peak will exhibit the asymmetry shown in figure 1. It is seen that the relative values of the resistance to mass transfer terms is controlled by the functions, $\frac{c_1 d_p^2}{D_m}$ and $\frac{c_2 d_f^2}{D_S}$.

It would seem that, in practice, the inequality defined in (3) can frequently occur but the converse does not appear to be true. Thus, peak asymmetry (in part or whole) resulting from inequality in mass transfer between the two phases manifests itself in the form shown in figure 1.

Alternatively, peak asymmetry could arise from thermal effects. During the passage of a solute along the column the heats of adsorption and desorption that are evolved and adsorbed as the solute distributes itself between the phases. At the front of the peak, where the solute is being continually adsorbed, the heat of adsorption will be *evolved* and thus the front of the peak will be at a temperature above its surroundings. Conversely, at the rear of the peak, where there will be a net desorption of solute, heat will be *adsorbed* and the temperature or the rear of the peak will fall below its surroundings.

Figure 2

Temperature Profile of a Peak Passing Through a Heat of Adsorption Detector

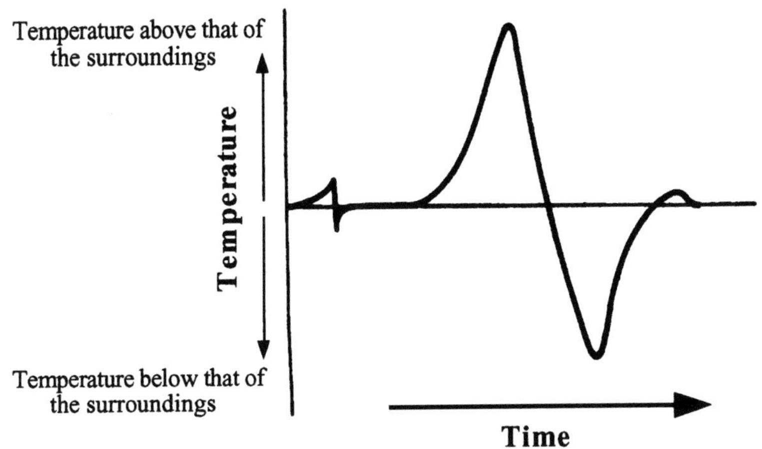

In fact, this phenomenon has been used as the basis of a very sensitive detecting system. An example of the temperature profile of an adsorbent as a peak passes over it is shown in figure 2. Unfortunately, it was found almost impossible to produce a true simulation of the *concentration* profile of the peak from the *temperature* profile and interest in the detector declined.

Nevertheless, it can be seen from figure 2 that the evolution and adsorption of heat will cause the front of the peak to travel through the column at an elevated temperature relative to the rear of the peak. Furthermore as the heat evolved will be proportional to the mass adsorbed, the largest temperature changes will occur at the highest concentrations in the peak. As a result, the distribution coefficient of the solute with respect to the stationary phase will be slightly smaller at the front of the peak and at the high concentrations, i.e. the peak maximum. Now, it has already been established that the velocity of a solute band through a column is inversely proportional to its distribution coefficient. Hence, the high concentrations of the peak-front will move slightly faster through the column than the lower concentrations of the peak. This will cause the peak maximum to move towards the front of the peak and thus produce peak asymmetry. The effect of the heat of adsorption on peak shape has been treated quantitatively for both LC (1) and GC (2).

In practice, it is probable that both of the effects discussed contribute to the overall peak asymmetry. Unfortunately, peak asymmetry varies in extent from the very obvious to the barely noticeable and because of this, peak asymmetry is often dismissed as the normal shape of a single solute peak. Such an assumption can cause serious errors in both qualitative and quantitative analysis.

Under circumstances where two solutes are incompletely resolved and one of the pair is present at a much lower concentration than the other, the profile of the pair often resembles a normal peak with slight asymmetry. Consider the combined elution profile of the two peaks shown in figure 3.

Figure 3

The Combined Elution Profile of Two Unresolved Peaks

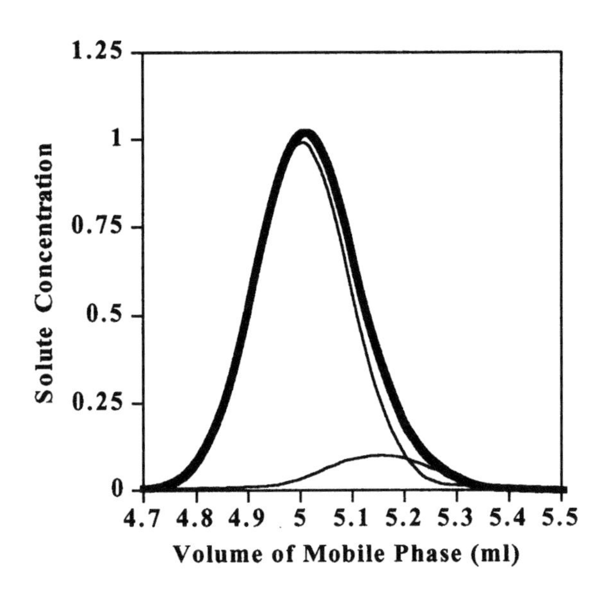

It is seen that the peak profile shown in figure 1, resulting from an asymmetric single peak, is indistinguishable from the asymmetric peak shown in figure 3 that results from a 10% impurity eluting relatively close to the parent peak. Care must be taken not to interpret peak asymmetry as the normal elution profile when, in fact, it is an indication of incomplete resolution. If there is any uncertainty, the solvent should be changed and the peak in question carefully monitored. It does not matter if, in changing the solvent mixture, the resolution of other components is lost, as only the integrity of the peak in question is being examined. The composition of the mobile phase should be changed at least twice to be sure that only a single solute is present. If no second component can be detected from the peak profiles, then the original mobile phase can be used for the analysis and the integrity of the peak assumed.

This an excellent example of the value of the diode array detector. If the chromatogram shown in figure 3 was monitored at two different wavelengths, then a peak ratio curve would immediately disclose the presence of the second peak (see page 175) and it would no longer be necessary to resort to changes in mobile phase composition to establish the presence of the impurity.

A second component, even at low concentration, will not only give an erroneous value for the amount of solute present but, on examining figure 3 it is seen that the retention time of the major peak is also significantly changed. This effect can be used for quantitative estimation of the unresolved mixture providing the chromatographic sytem can provide the necessary high precision.

Scott and Reese (3) measured the accuracy and precision that could be obtained from a well designed liquid chromatograph and their work will be referred to again later. In the course of their work they found that, providing the retention time of a mixed peak could be measured with sufficient accuracy and the retention times of the two pure components were known, then the proportion of the two components could be determined, even if only a single peak was apparent. The theory behind this work is a little complex (4), and it is not appropriate to discuss the details here. The basis of the method was to calculate the theoretical retention time from the combined elution curves for the two solutes over a range of mixtures. The curve relating retention time to composition was then used as a calibration curve to determine the composition of an unknown mixture from the retention time of the combined peak. The calibration curve that was calculated and the experimental points they obtained are shown in figure 4.

It is seen that the solutes used were nitrobenzene and deutero-nitrobenzene. These two solutes are so similar in physical and chemical characteristics that neither can be exclusively detected in a solvent without interference by the other. It is also seen that the retention times of the two solutes are 8.927 minutes and 9.061 minutes giving a difference of only 8.04 sec.

Figure 4

Graph of Retention Time Difference against Sample Composition for Mixtures of Nitrobenzene and Deutero-Nitrobenzene

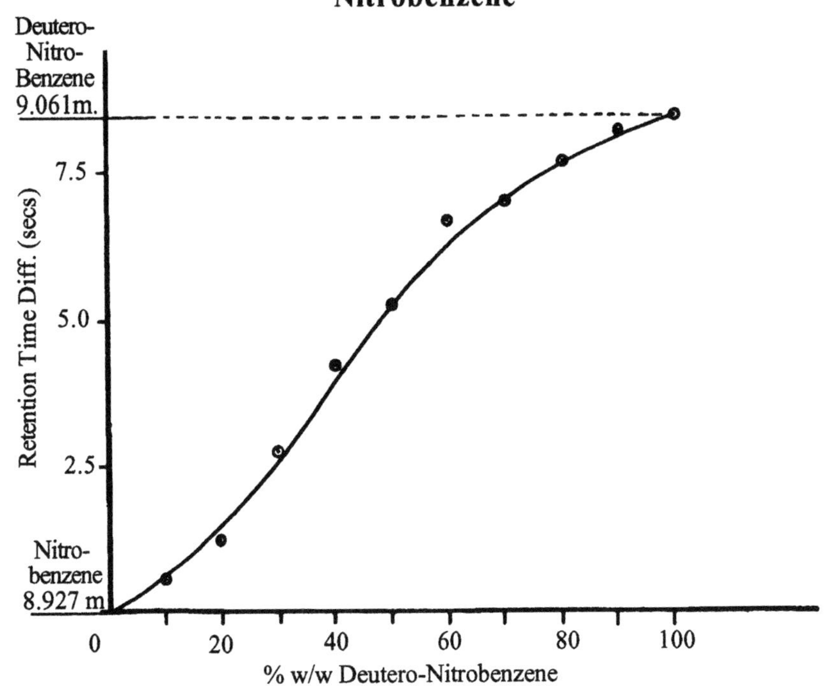

It follows that measurements must be made with a precision of about 0.2 second if quantitative results are to be of any value. It is seen from figure 4 that the experimental points lie very close to the line and a fairly accurate measurement of the distribution of the two isotopes can be obtained from retention time measurements. This method has very limited areas of application and is given here, more to demonstrate the effect of unresolved impurities on retention time, than to suggest it as an alternative to adequate chromatographic resolution. In some cases, however, particularly in the analysis of isotopes, it may be the only practical way to obtain a quantitative evaluation of the mixture by a liquid chromatographic method.

Finally, it must be emphasized that the uncertainty arising from the slight asymmetry shown by the presence of a 10% impurity in figure 3 does not depend on the magnitude of the impurity. In figure 5 the elution curve resulting from a 50% mixture of two closely eluting components is shown.

Figure 5

The Elution Curve of a Binary Mixture Containing Equal Quantities of Each Solute

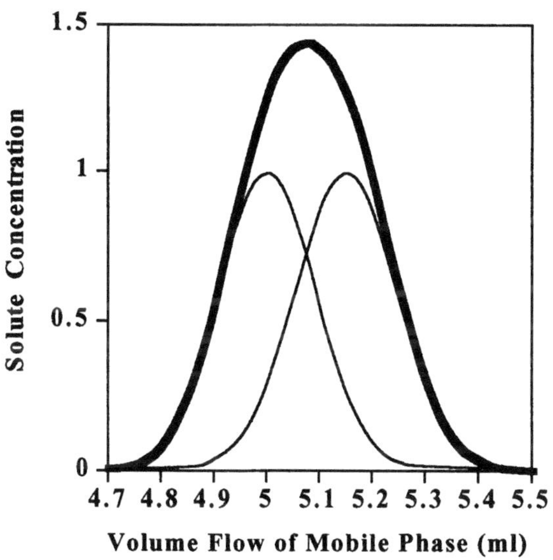

It is seen that the profile of the combined peaks is perfectly symmetrical and displays no hint that there are two solutes present. Obviously an absorption ratio curve from a diode array detector would quickly disclose the presence of the two components, as would an appropriate changes in mobile phase composition. However, there would be a further clue for the analyst to follow that would give warning of the "duplicity" of the peak. The double peak would be very broad and be inconsistent with the change in peak width of the other solute peaks with retention time. The peak width of a solute increases regularly with retention time but, unfortunately, the relationship is not smooth. There are good reasons for this, but they are not pertinent to this discussion. However, a peak such as that shown in figure 5 would be excessively broad and obviously out of order with the peak widths of neighboring solutes. This should alert the analyst to the possibility of an unresolved pair of solutes. The analyst should always be alert to the width of each peak in relationship to its position on the chromatogram, as this is the first indication of the presence of a composite peak.

The Effect of Temperature on Retention Volume Measurement

There are two major factors that influence retention volume measurement and they are temperature and solvent composition. In order to measure retention volume with adequate precision it is necessary to know the relationship between retention time and temperature so that the control limits of the column temperature can be specified.

The effect of temperature on retention time was investigated by Scott and Reese (3), who measured the retention volume of the solutes o-dinitro-benzene, 2-ethoxy naphthalene and p-chlorophenatole over a range of temperatures. The chromatographic conditions used are as follows,

Chromatographic Conditions

Column	Silerex 1
Column Length	25 cm
Column Diameter	4.6 mm
Column Packing	Silica Gel (particle size 10μm)
Mobile Phase	44% butyl chloride and 56% n-heptane
Flow Rate	1 ml/min.
Detector	UV adsorption at 254 nm
Sample Volume	1 μl

Scott and Reese chose to monitor retention volume as opposed to retention time, as retention volume is always the primary dependent variable in LC. Retention time is not a primary measurement because it must also include the reproducibility of the flow-rate delivered by the pump. The retention volume was measured absolutely by means of a burette connected to the end of the column. The column was situated in a water bath that was controlled to +/- 0.05ºC. Precautions were taken to ensure the mobile phase attained the column temperature before entering the column. The results obtained are shown in figure 6A.

Figure 6A

Graphs of Retention Volume against Temperature for Three Solutes

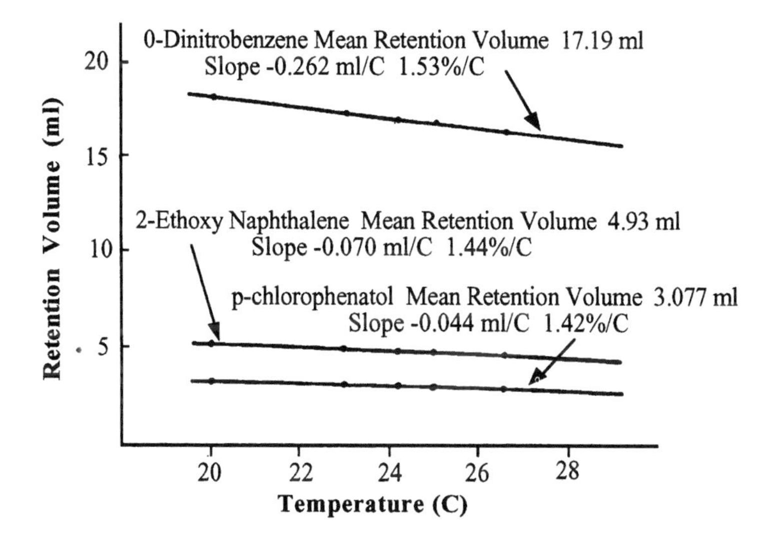

It is seen that in order to measure retention volumes with a precision of 0.1%, the temperature control must be +/- 0.04ºC. This level of temperature control on a thermostat bath is not difficult to achieve but it is extremely difficult, if not impossible, to return to a specific temperature to within +/- 0.04ºC after prior change. To achieve a precision of retention volume measurement of 1%, the temperature control must be +/- 0.4ºC. This is far more practical as most column oven temperature can be set to a given temperature to within +/- 0.25ºC. Although the data was obtained for three specific solutes, the results can be taken as reasonably representative for all solutes and phase systems. In most practical analyses, the precision limits of retention volume measurement will be about 1% but this will not include the reproducibility of the flow rate given by the pump. As retention *times* are usually measured in analytical LC, the precision of measurement may be significantly greater than +/-1%. It follows that if the identity of the peak must be confirmed, and retention data is being used for the purpose, then it is essential that the column is carefully thermostatted.

The Effect of Solvent Composition on Retention Volume Measurement

Scott and Reese (3) also measured the change in retention volume with solvent composition using the same LC apparatus as that used for investigating the effect of temperature. The column was thermostatted at 24.7°C and the results that were obtained are shown in figure 6B.

Figure 6B

Graphs of Retention Volume against Solvent Composition for Three Solutes

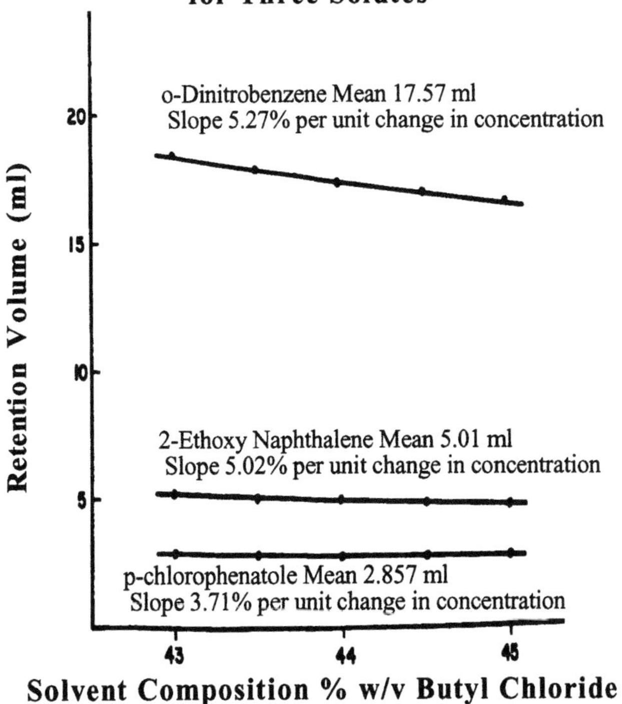

It is seen that small changes in solute composition can have a profound effect on the retention volume, but it must be borne in mind that the magnitude of the effect will vary somewhat, between different solutes and different phase systems. The results shown in figure 6B indicate that if a retention volume is to be measured with a precision of 0.1%, the solvent composition must be maintained constant to within 0.02% w/v. Normally, the solvent composition can be kept constant within those limits providing a closed solvent system is used to prevent evaporation. However, it would be extremely difficult, if not impossible to make up another solvent mixture of the same composition within those limits of precision. This would be particularly difficult if either of the solvents were volatile. If the precision required for retention volume measurement was 1%, the solvent composition would have to be maintained constant with a precision of +/- 0.2 % w/v which should be quite practical. Furthermore, it would be fairly straightforward to make up replacement solvents to the same concentration within the same precision limits. For a routine analysis, however, it might well be advantageous to prepare the mobile phase in large volumes and store it in an appropriate manner. Scott and Reese (3) examined the repeatability of retention time measurements by two procedures, one employing computer processed data and the other by manual measurement of the distance in centimeters from the injection point to the peak maximum on a potentiometric recorder chart. They carried out twelve replicate analyses and the resulting statistical analysis is shown in table 1.

Table 1

The Precision of Retention Data Measurement Made Manually and by Computer

Parameter	Peak 1	Peak 2	Peak 3	Peak 4
Capacity factor	0.22	0.94	1.50	5.21
Mean (cm)	10.17	16.0	20.59	51.14
S.D.(% Mean)	0.85	0.245	0.19	0.15
Mean (min.)	3.97	6.27	8.11	20.14
S.D.(% Mean)	0.31	0.20	0.17	0.33

It is a little surprising to see from table 1 that there is only a minor difference between the precision of the two methods of measurement. This might suggest that the difference between replicate values was not arising from the methods of measurement but was caused by variations elsewhere in the chromatographic system. In order to gain some insight into the source of the variations in retention data, Scott and Reese (3) reconstructed the tips of the two peaks that gave the smallest and largest value of the twelve set of replicates for peak 4. The two peaks are shown in figure 7.

Figure 7

Tips of Peaks Having Extreme Retention Values Reconstructed by the Computer

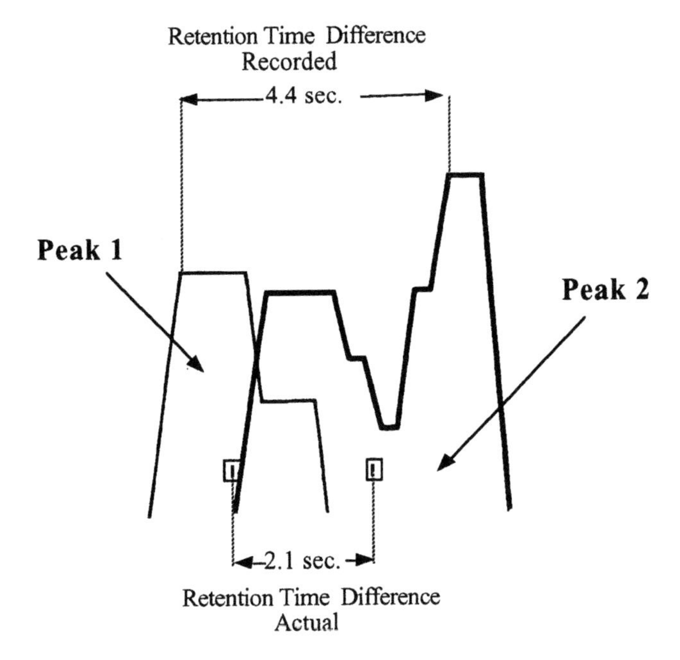

By taking the mean positions of the peaks, it is seen that they are only 2.1 sec apart. However, as a result of a noise spike on the front of the first peak and a similar noise spike on the back of the second peak, the computer interpreted each noise spike as the peak maxima, which resulted in a retention difference of 4.4 sec. It is clear that a significant part of the scatter of retention data can be due to detector noise, either arising from the detector itself or from changes in operating conditions of the chromatograph. This once more emphasizes the importance of using a chromatograph that can maintain stable operating conditions in conjunction with a detector having as little long term noise as possible. The results also indicate that unless the maximum sensitivity is essential for the analysis, the sample size should be adjusted so that the detector can be operated well below its maximum sensitivity to reduce the consequence of the noise.

Quantitative Analysis

Quantitative estimates of the mass of a particular solute present in a sample are obtained from either peak height or peak area measurements. The values obtained are then compared with the peak height or area of a reference solute present in the sample at a known concentration or mass. In this chapter quantitative analysis by LC will be discussed but the procedures described should not be considered as entirely appropriate for other types of chromatographic analysis. Those interested in general quantitative chromatographic analysis including GC and TLC are referred to the book by Katz (4).

Peak Height Measurements

Most detectors are concentration sensitive devices and thus the peak height will be proportional to the maximum concentration in the peak, which, in turn, will be proportional to the total area of the peak. The total area of the peak is proportional to the total mass of solute contained in the peak providing it is not excessively tailing. As the peak height is inversely related to the peak width, then, if peak heights are to be used for analytical purposes, all parameters that can affect the peak width must be held constant. This means that the capacity factor of the solute (k') must remain constant and, consequently, the solvent composition is held stable. The temperature must also be held constant and an highly repeatable method of sample injection must be used. If computer data acquisition and processing are employed, then a direct printout of the peak heights is obtained and, with most systems, the calculated analysis is also presented. If the peak heights are to be measured manually, which even today is carried out in the majority of LC analyses, then the base line is produced beneath the peak and the height between the extended base line and the peak maximum measured. In general, the measurements should be made estimating to the nearest 0.1 mm.

Peak Area Measurements

The area of a peak is the integration of the peak height (concentration) with respect to time (volume flow of mobile phase) and thus is proportional to the total mass of solute eluted. Measurement of peak area accommodates peak asymmetry and even peak tailing without compromising the simple relationship between peak area and mass. Consequently, peak area measurements give more accurate results under conditions where the chromatography is not perfect and the peak profiles not truly Gaussian or Poisson.

Unfortunately, neither the computer nor the potentiometric recorder measures the primary variable, *volume of mobile phase*, but does measure the secondary variable, *time*. This places stringent demands on the LC pump as the necessary accurate and proportional relationship between time and volume flow depends on a constant flow rate. Thus, peak area measurements should never be made unless a good quality pump is used to control the mobile phase flow rate. Furthermore, the pump must be a constant flow pump and *not* a constant pressure pump.

Peak areas can be measured manually in a number of ways, the simplest being the product of the peak height and the peak width at 0.6065 of the peak height. This does not give the true peak area but providing the peak is Poisson, Gaussian or close to Gaussian it will always give accurately the same proportion of the peak area. Other methods involve the use of a planimeter (an instrument that provides a numerical value for the area contained within a perimeter traced out by a stylus), and the measurement of the product of the peak height and the peak width at the base. The first method is extremely tedious and the second somewhat inaccurate. The most accurate manual method of measuring peak area is to cut the peak out and weigh it. A copy of the chromatogram should be taken and the peaks cut out of the copy. This procedure is also a little tedious (not as tedious as the use of a planimeter) but does provide very accurate values for the peak area. It is particularly effective for skewed or malformed peaks where other methods of manual peak area measurement (with the exception of the planimeter) fail dismally and give very inaccurate results. The recommended method is to use the product of the peak height and the peak width at 0.6065 of the peak height but this does require adequate resolution of the components of the mixture.

Procedures for Quantitative Analysis

There are two basic methods used in quantitative analysis; one uses a reference standard with which the peak areas (peak heights) of the other solutes in the sample are compared; the other is a normalization procedure where the area (height) of any one peak is expressed as a percentage of the total area (heights) of all the peaks. There are certain circumstances where each method is advantageous, and providing they are used carefully and appropriately all give approximately the same accuracy and precision.

Quantitative Analysis Using Reference Standards

Reference standards can be used in two ways; a weighed amount of the standard can be added directly to the sample and the area of the peaks of interest compared with that of the standard; alternatively, a weighed amount of the standard can be made up in a known volume of solvent, a sample placed on the column and chromatographed under exactly the same conditions as the original sample. The peak area (height) of the reference solute in the reference chromatogram is compared to the peak areas (heights) of the solutes of interest in the sample chromatogram. The details of the two methods will now be described separately.

The Internal Standard Method

The use of an internal standard probably gives the most accurate quantitative results. However, the procedure depends upon finding an appropriate substance that will elute in a position on the chromatogram where it will not interfere or merge with any of the natural components of the mixture. If the sample contains numerous components, this may be difficult. Having identified a reference standard, the response factors for each component of interest in the mixture to be analyzed must be determined. A synthetic mixture is made up containing known concentrations of each of the components of interest and the standard. If there are (n) components, and the (r) component is present at concentration (C_r) and the standard at a concentration (C_{st}).

Then,

$$\frac{C_r}{C_{st}} = \frac{A_r}{A_{st}} \alpha_r$$

where (A_r) is the area of the peak for component (r),
(A_{st}) is the area of the peak for the standard,
and (α_r) is the response factor for component (r).

Thus, the response factor (α_r) for the component (r) is given by

$$\alpha_r = \frac{C_r A_{st}}{C_{st} A_r}$$

If peak heights are used instead of peak areas then

$$\frac{C_r}{C_{st}} = \frac{H_r}{H_{st}}\alpha_r$$

where (H_r) is the height of the peak for component (r),
(H_{st}) is the height of the peak for the standard,
and (α_r) is the response factor for component (r).

Thus, the response factor (α_r) for the component (r) is then given by

$$\alpha_r = \frac{C_r H_{st}}{C_{st} H_r}$$

A weighed amount of standard is now added to the sample so that its concentration is now $(C_{st(s)})$ and the sample chromatographed. If the concentration of any component (p) is (C_p) and the peak areas of the component (p) and that of the standard are (A_p) and $(A_{st(s)})$ respectively then

$$C_p = \frac{A_p C_{st(s)}}{A_{st(s)}}\alpha_p$$

Again if peak heights are employed instead of peak areas,

$$C_p = \frac{H_p C_{st(s)}}{H_{st(s)}}\alpha_p$$

where (H_p) is the peak height of component (p) in the sample,
and $(H_{st(s)})$ is the peak height of the standard in the sample.

Thus, the concentration of any (or all) of the components present in the mixture can be determined providing they all adequately separated from one another. If the same type of mixture is being analyzed, the operating conditions are maintained constant and there is not an extreme change in the composition of the samples, the response factors usually need be determined only once a day.

The External Standard Method

The external standard method requires that the standard is chromatographed separately from the sample and thus, the chromatographic conditions must be maintained extremely constant. The great advantage of the external standard method is that the reference standard (or standards) can be identical to the solute (or solutes) of interest in the sample. Thus, a synthetic mixture can be made up in which the concentration of the components is closely similar to those of the sample.

Thus, if the (p)th solute in the mixture is at a concentration of ($C_{p(s)}$) in the sample and ($C_{p(st)}$) in the standard solution, then

$$C_{p(s)} = \frac{A_{p(s)}}{A_{p(st)}} C_{p(st)}$$

where ($A_{p(s)}$) is the area of the peak for solute (p) in the sample chromatogram,
($A_{p(st)}$) is the area of the peak for solute (p) in the reference chromatogram,
and ($C_{p(st)}$) is the concentration of the standard in the reference solution.

If peak heights are employed, then

$$C_{p(s)} = \frac{H_{p(s)}}{H_{p(st)}} C_{p(st)}$$

where ($H_{p(s)}$) is the area of the peak for solute (p) in the sample chromatogram,
($H_{p(st)}$) is the area of the peak for solute (p) in the reference chromatogram,
and ($C_{p(st)}$) is the concentration of the standard in the reference solution.

Theoretically, providing the chromatographic conditions are kept constant, then, the reference chromatogram need only be run once a day. However, it is advisable to run the reference chromatogram at least every two hours and many analysts run a reference chromatogram immediately after each sample.

The Normalization Method

The normalization method is the easiest and most straightforward to use but, unfortunately, it is also the least likely to be appropriate for most LC analyses. To be applicable, the detector must have the same response to all the components of the sample. An exceptional example, where the normalization procedure is frequently used, is in the analysis of polymers by exclusion chromatography using the refractive index detector. The refractive index of a specific polymer is a constant for all polymers of that type having more than 6 monomer units. Under these conditions normalization is the obvious quantitative method to use.

The percentage x(p)% of any specific polymer (p) in a given polymer mixture can be expressed by

$$x_p = \frac{A_p}{A_1 + A_2 + A_3 + \ldots + A_n} 100$$

where (A_p) is the peak area of polymer (p)

or $$x_p = \frac{A_p}{\sum_{p=1}^{p=n} A_p} 100$$

If peak heights are used, the percentage x(p)% of any specific polymer (p) in a given polymer mixture can be expressed by similar equations,

$$x_p = \frac{H_p}{H_1 + H_2 + H_3 + .\ .\ .\ . + A_n} 100$$

where (H_p) is the peak area of polymer (p)

or

$$x_p = \frac{H_p}{\sum_{p=1}^{p=n} H_p} 100$$

The Relative Precision of Peak Height and Peak Area Measurements

In their work on the precision of contemporary liquid chromatographic measurements, Scott and Reese (3) also evaluated the precision that could be expected from a computer measuring peak heights and peak areas. They again used twelve replicate samples and the results they obtained are shown in table 2.

Table 2

Precision Obtained from the Measurements of Peak Heights and Peak Areas by Means of a Computer

Parameter	**Peak 1**	**Peak 2**	**Peak 3**
Capacity Ratio (k')	0.94	1.50	5.21
Peak Heights			
Mean	1785	15191	75140
Stand.Dev. (S.D.)	63	319	1183
S.D.(% of Mean)	3.5	2.1	1.6
Peak Areas			
Mean	282929	3344011	41042092
Stand.Dev. (S.D.)	17301	64503	583521
S.D.(% of Mean)	6.11	1.93	1.42

It is seen that there is not a great difference between the use of peak heights or peak areas for quantitative analysis, except possibly for very early peaks, where the results seem to indicate that peak height measurements might be more precise. However, it again must be emphasized that the measurements made by Scott and Reese were *overall* precision measurements that will include all variations in the chromatographic system. The difference between the two methods of measurement may well be significant, but the absolute values for precision will not, by any means, be solely dependent on the method of peak measurement.

Peak De-Convolution

In some data acquisition and processing systems, software is included that mathematically analyze convoluted (unresolved) peaks, identify the individual peaks that make up the composite envelope and then determine the area of the individual peaks.

It must be stressed at this point that clever algorithms are no substitute for good chromatography.

Sometimes it is not possible to improve the resolution of a complex mixture beyond a certain level and, under these circumstances, the use of some de-convolution technique may be the only solution. The algorithms in the software must contain certain tentative assumptions in order to analyze the peak envelope. Firstly, a particular mathematical function must be assumed that describes the peaks. The function used is usually Gaussian and, in most cases, no account is taken of the possibility of asymmetric peaks. Furthermore it is also assumed that *all* the peaks can be described by the *same* function (i.e. the efficiency of all the peaks are the same) which, as has already been discussed, is also not generally true. Nevertheless, providing the composite peak is not too complex, de-convolution can be reasonably successful.

If resolution is partial and the components are present in equal quantities, then the de-convolution approach might be appropriate. An example of the convolution of two partially resolved peaks of equal size is shown in figure 8.

Figure 8

The De-Convolution of Two Partially Resolved Peaks Representing Solutes Present in Equal Quantities

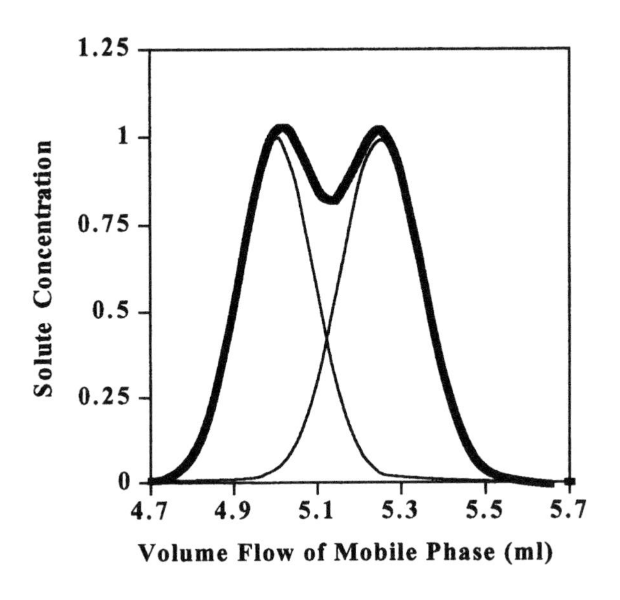

It is seen that two Gaussian shaped peaks can be easily extracted from the composite envelope and the software could also supply values for either the heights of the de-convoluted peaks or their area. It should be noted, however, that the two peaks are clearly discernible as unresolved components in the original chromatogram. As a consequence, the de-convoluting software can easily identify the approximate positions of the peak maxima and can also obtain a crude value for the peak widths. Such information allows the software to arrive at a valid analysis quickly and with reasonable accuracy. Nevertheless, it should also be pointed out, that in the example shown in figure 8 most data processing software would construct a perpendicular from the valley to the base line, thus bisecting the combined peak envelope. The area of each half would then be taken as the respective area of each peak.

The software will also function well if the same partial resolution is obtained but for peaks of unequal size. An example of the analysis of such a composite peak is shown in figure 9.

Figure 9

The De-Convolution of Two Partially Resolved Peaks Representing Solutes Present in Unequal Quantities

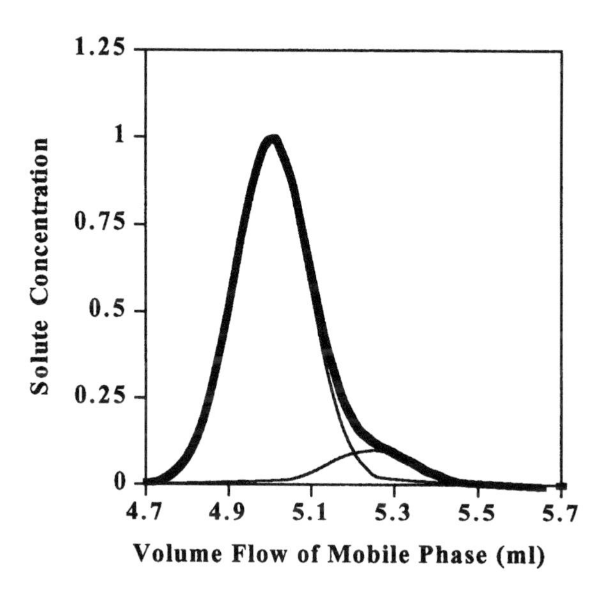

It seen that the de-convolution is likely to be successful as the position of the peak maximum, and the peak width, of the major component is easily identifiable. This would mean that the software could accurately determine the constants in the Gaussian equation that would describe the profile of the major component. The profile of the major component would then be subtracted from the total composite peak leaving the small peak as difference value. This description oversimplifies the calculation processes which will include a number of iteration steps to arrive at the closest fit for the two peaks.

However, in the examples taken the solutes were at least partly resolved and, unfortunately, as the resolution becomes less and less and the need for an accurate de-convolution technique becomes even greater, the value of the software presently available appears to become minimal.

A typical example at the other extreme where a de-convolution technique would be useless is given in figure 10.

Figure 10

The De-Convolution of Three Completely Unresolved Peaks

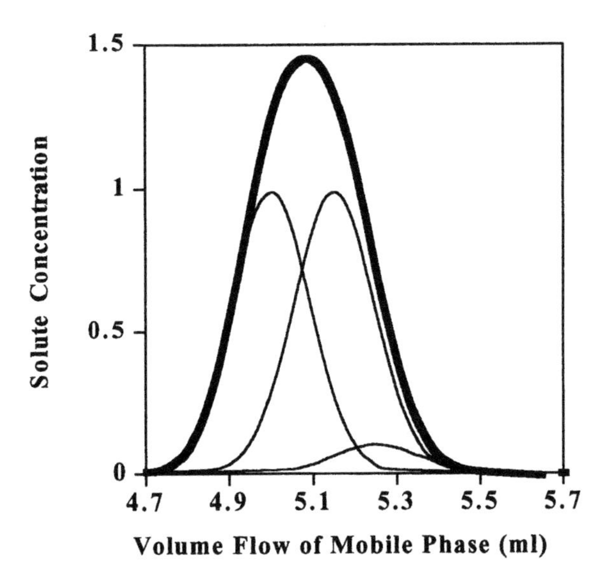

It is seen that however sophisticated the software might be, it would be virtually impossible to de-convolute the peak into the three components. The peaks shown in the diagram are discernible because the peaks themselves were assumed and the composite envelope calculated. The envelope, however, would provide no basic data; there is no hint of an approximate position for any peak maximum and absolutely no indication of the peak width of any of the components. The use of the diode array detector, monitoring at different wavelengths, might help by identifying uniquely one or more of the components. If the software could import extra data of this type then de-convolution might be possible. The use of de-convoluting software has limited application and must be used with considerable caution. Every effort should be made to separate all the components of a mixture by chromatography and only employ peak de-convolution as a last resort.

Reporting Analytical Results

Modern data processing software varies widely in both the data that is provided and the manner in which it is presented. Most systems give a choice of peak height, or peak area measurements, or both, which provide a routine for determining the response factors for each solute from the calibrating chromatogram. Retention times are taken as the time at which the peak maximum is reached and if the dead volume peak is defined, then the capacity ratios of the components will also be calculated. From the capacity factors a simple ratio algorithm can provide the retention ratio of each component to any previously defined solute peak. Some data systems provide peak-valley identification and also peak-valley-to-baseline construction. A peak valley is the minimum between two incompletely resolved peaks. Peak skimming routines are also common which are used when a small peak is eluted on a badly tailing major peak. The tail of the peak is constructed (actually extended) under the small peak and the area of the small peak taken as that peak area above the extended tail.

The information provided on the print-out also varies considerably from one manufacturer to another. Some provide all the chromatographic information, including solvent composition, the gradient program if one is used, column type, dimensions and operating temperature together with the mobile phase flow rate. In some cases the type of detector detector operating conditions and the sensitivity (or attenuation) setting is reported. In general, all the above information is normally reported with the analysis and so it is very convenient to have the information included in the computer print-out.

Figure 11 Computer Print-Out of an LC Analysis

Sample Volume 1 μl Dilution 1 to 1

External Standard Area 10000 External Standard Concentration 1000 fmole

Retention Time	Peak Area	Peak Height	Concentration fmol
0.363	33318956	7439.579	-
3.590	206343	34.884	-
4.430	16853	1.541	-
5.150	27775	1.208	27775

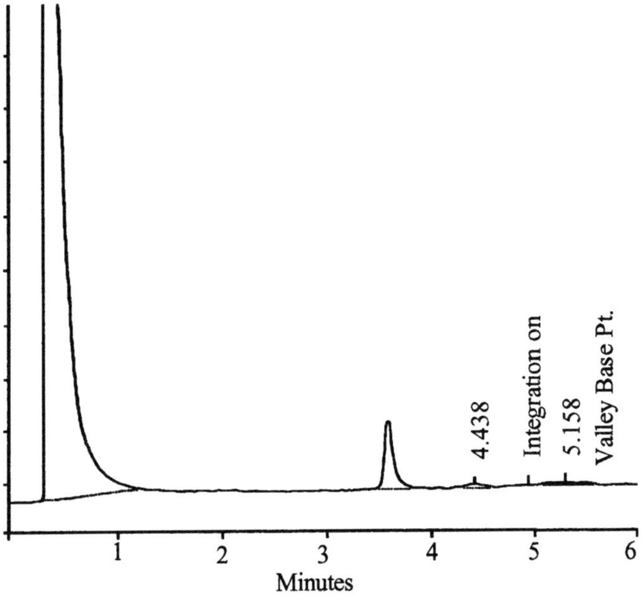

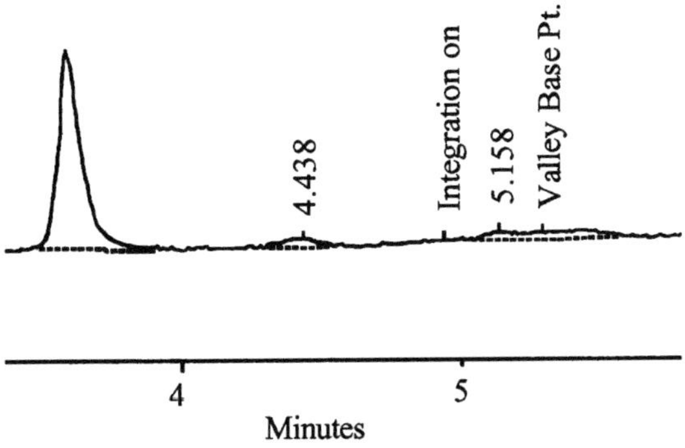

As one might expect, computer data processing is an expensive adjunct to a liquid chromatograph and the more sophisticated reporting systems can be very expensive. An example of a computer print-out of

a trace analysis is shown in figure 11. The example that is taken is not as neat and explicit as those usually shown in the literature but more typical of those that the analyst is likely to meet in day-to-day analyses. The sample is of biological origin and the material of interest is present in trace quantities. The area of the chromatogram depicting the separation of the substance being determined is shown enlarged at the base of the diagram. This facility for enlarging a chosen portion of the chromatogram, for detailed inspection, is also a common feature of many data processing systems. It is seen that the substance of interest, that eluted at a retention time of 5.158 min., is incompletely resolved from its neighbor which is eluted later. It is also seen that the software marks where integration is initiated and where it ends at the valley-base line. The analyst can, therefore, check the limits of the integration and thus, be assured that only the area of the peak of interest was being calculated.

Although computer systems have become relatively inexpensive and the software for processing chromatography data is fairly straightforward to write, the data acquisition and processing systems (albeit highly desirable) are still a significant added cost to the overall chromatograph. As a result, even today, manual measurements made directly on the chromatogram are still used in a large proportion of all LC analyses. Results obtained by manual measurements suffer from little loss, if any, in accuracy and precision but are certainly time consuming. If LC techniques are being used for quality control analyses, where a large number of samples are chromatographed every day, then some form of data acquisition and processing facilities are essential.

References

1/ R.P.W. Scott, *"Liquid Chromatography Theory"*, John Wiley and Sons,Chichester-NewYork-Brisbane-Toronto-Singapore, (1992)77.

2/ R.P.W. Scott, *Analytical Chemistry,* **35**(1963)481.

3/ R.P.W. Scott and C.E. Reese, *J. Chromatogr.*, **138**(1977)283.

4/ *"Quantitative Analysis using Chromatographic Techniques"* (Ed. E. Katz), John Wiley and Sons, Chichester-NewYork-Brisbane-Toronto-Singapore (1987).

9
LC Applications

The purpose of this final chapter is to provide the analyst with a background of practical examples to aid in the selection of, firstly, the best chromatographic method and, secondly, the best phase system when faced with an hitherto unknown sample for analysis. The literature is rich with LC applications and frequently publications are available for the separation of closely similar mixtures to that of the sample. It is unlikely, however, that the chromatographic conditions for the actual separation required will be available and, even if they are, the conditions reported may well not be optimum. This is more likely to be true for those applications that are described in earlier publications. Nevertheless, conditions that have be successfully employed for related separations may certainly help to identify those conditions necessary for the sample supplied for assay.

It is important that chromatographic conditions reported in the literature should not be imitated "parrot fashion" and that the separation is examined logically in the light of the retentive mechanisms that are likely to be involved. Changes should be made on a rational basis, grounded on an understanding of the chromatographic system that is employed (e.g., exclusion chromatography, ion exchange chromatography, etc.), and of the type of interactions that are likely to be involved (e.g., dispersive, polar, ionic, etc.). A number of different analyses will be examined and used to illustrate the rationale behind the choice of the separation method and its associated phase system. This will direct the reader to the same logical approach in selecting the appropriate chromatographic parameters for his/her separation problem.

The most important chromatographic systems that are available to the analyst are as follows:

1/ Exclusion systems

2/ Chiral systems

3/ Interactive systems

a/ Dispersive
b/ Polar
c/ Ionic
d/ Combinations of dispersive, polar and ionic

Examples of each of the above chromatographic systems have, for the most part, been chosen *deliberately* from *manufacturers data sheets*. As a consequence, complete details of the separation will be available to the analyst from the respective manufacturer, if needed, and furthermore, the actual stationary phase can also be supplied if required.

Separations Based on Exclusion Chromatography

It should be recalled that all substances that are used as stationary phases, or as supports for bonded phases, that have pores commensurate with the size of the molecules being separated, will exhibit exclusion properties. Thus, even if the solutes are retained largely as a result of the interactions of the solute molecules and those of the two phases if, due to their size, some molecules can interact with more stationary phase than others, then the retention will also be controlled to some extent by exclusion. The term *exclusion chromatography* is, therefore, usually confined to those separations where retention is controlled almost *solely* by exclusion, molecular interaction of the solute molecules with the two phases being almost identical.

Exclusion chromatography is usually employed to separate mixtures of polymers where the interactive properties of each solute are very similar and the major difference between them is molecular size. Examples would be mixtures of polystyrenes, or polyethylene glycols. Another area where exclusion chromatography is useful is where there is a multitude of interactive sites on the solute molecules that also results in the interactive capacity of each of the solutes being very similar. This can occur if the molecules are made up of large numbers of common fragments, for example high molecular weight proteins comprised of a large number of amino acids. The dispersive, polar and ionic interactive sites of each solute resulting from the multitude of common amino acids can, under certain circumstances, be very similar. Consequently, there will be little or no differential molecular interactions between the different solutes, irrespective of the phase system selected. Under these circumstances, separation on the basis of molecular size may be the only practical alternative.

There are two types of stationary phases commonly used in exclusion chromatography; silica gel and micro-reticulated cross-linked polystyrene gels. A third type of exclusion media is comprised of the Dextran gels. Dextran gels are produced by the action of certain bacteria on a sucrose substrate. They consist of framework of glucose units that can form a gel in aqueous solvents that have size exclusion properties. Unfortunately the gels are mechanically weak and thus, cannot tolerate the high pressures necessary for HPLC and, as a consequence, are of very limited use to the analyst.

Exclusion Chromatography Employing Silica Gel

Silica gel was the first stationary phase to be used effectively at high pressures for exclusion chromatography. Silica gel is mechanically strong and, as already discussed, can be made available with a wide range of pore sizes. It is particularly useful in the separation of polymers that are either weakly polar or dispersive in character. Consequently, if used with a polar solvent, the silica gel can be deactivated by an adsorbed layer of solvent on the surface so that there is no net relative interaction between the solute and the two phases. Under such circumstances, the separation is entirely due to size exclusion. An example of the use of silica gel for a separation by size exclusion is shown in figure 1.

Figure 1

The Separation of Some Polystyrene Standards by Exclusion Chromatography on Silica Gel

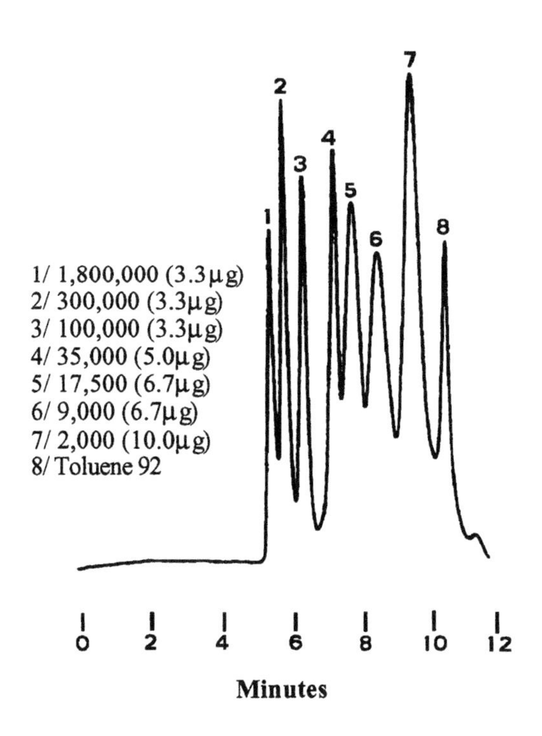

Courtesy of Supelco Inc.

All the solutes in exclusion chromatography are eluted between the interstitial column volume and the column dead volume (i.e. the pore volume). Consequently, the column must be large enough to provide adequate pore volume to contain all the peaks and at the same time, maintain narrow solute bands by achieving high efficiencies in order to achieve the desired resolution. The simple theory described previously in this book can be used to examine the separation that is shown in figure 1.

The column used in the separation depicted in figure 1 was 25 cm long and 6.2 mm in diameter packed with silica gel having a mean pore diameter of 100 Å and a particle diameter of 5 μm. Thus, the column would have an HETP of approximately 0.001 cm (twice the particle diameter). Consequently, a column 25 cm long would have an efficiency of about $\frac{25}{.001} = 25{,}000$ theoretical plates. It is seen from figure 1 that the dead time is about 12 ml, which at the reported flow-rate of 1 ml/min gives a dead volume of 12 ml. The peaks were eluted between about 5 minutes and 12 minutes so it would appear that the pore volume (the volume containing the peaks) would be 7x1 = 7 ml. Bearing in mind that the peak volume is given approximately by

$$\frac{4V_o}{\sqrt{n}} = \frac{4x12}{\sqrt{25{,}000}} = 0.481 \text{ ml}$$

Thus, the peak capacity is given by

$$\frac{\text{Pore Volume}}{\text{Peak Volume}} = \frac{7}{0.481} = 14$$

Thus, the column should completely resolve about 14 equally spaced peaks. It is seen from figure 1 that a peak capacity of 14 is not realized although most of the components are separated. This means that the column may not have been packed particularly well and/or the flow rate used was significantly above the optimum velocity that would provide the maximum efficiency. The mobile phase that was used was tetrahydrofuran which was sufficiently polar to deactivate the silica gel with a layer (or perhaps bilayer) of adsorbed solvent molecules yet was sufficiently dispersive to provide adequate sample solubility. This simple arithmetical exercise demonstrates the need to have well-packed columns and to operate at the optimum velocity if the maximum efficiency is required. It also shows that the efficiency of any column purchased should be measured on delivery to ascertain the quality of the packing and if it is adequate for the intended separation.

Unfortunately, silica gel suffers from a severe disadvantage in that it is slightly soluble in water. This means that the native silica cannot be used in conjunction with aqueous mobile phases. Consequently, silica is precluded from use in the separation of those substances that are strongly polar and require aqueous solvents to render them soluble and stable. For this reason the relatively new, micro-reticular resins are now used in the separation of strongly polar substances by exclusion chromatography.

Exclusion Chromatography Employing Micro-Reticulated Cross-Linked Polystyrene Gels

Micro-reticulated polystyrene gels, as already discussed, are made from cross-linked styrene-divinyl benzene polymers and can be produced with a wide selection of pore sizes. However, they are not used solely for the separation of large multi-interactive molecules (such as certain protein mixtures), but can be used in the same manner and for the same type of mixtures as silica gel. The pore volume of the gels, however, due to their method of manufacture, may be significantly lower than that of silica gel. Consequently, some gels may have neither the peak capacity nor the loading capacity normally experienced with silica gel but, this is generally not a problem in analytical LC. An example of the separation of some phthalate esters is shown in figure 2 and demonstrates the use of a micro-reticulated gel in separating the type of mixture that could be also resolved on silica gel. The column was 30 cm long and 7.8 mm diameter. As noted, in exclusion chromatography a large column is necessary to provide adequate peak capacity. The particle diameter of the packing was 6 μm and thus, at the optimum velocity should provide an HETP of .0012 cm and consequently, an efficiency of about $\frac{30}{.0012} = 25{,}000$ theoretical plates. It is seen in figure 2 that the dead volume time appears to be about 19 minutes and at a stated flow rate of 1.0 ml/min would be equivalent to a dead volume of 19.5 ml.

Figure 2

The Separation of Some Phthalate Esters by Exclusion Chromatography on Styrene-Divinyl Benzene Based Gel

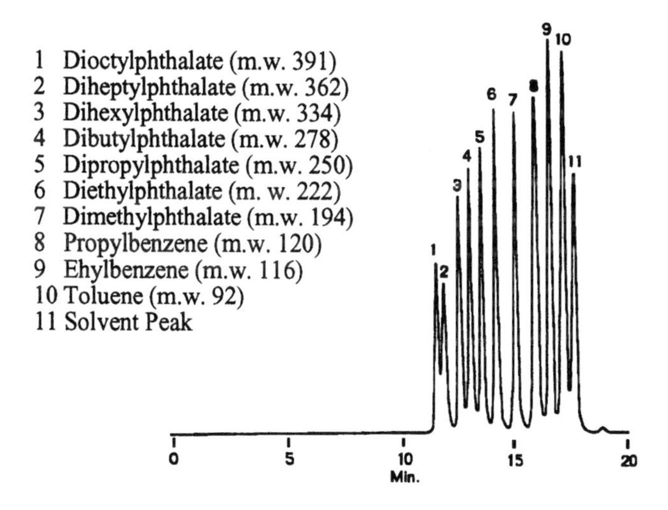

Courtesy of Supelco Inc.

It is also seen in figure 2 that the solutes are eluted between about 11 minutes and 19 minutes. Consequently the pore volume will be about (19-11)1.0 = 8 ml.

Now the average peak volume will be approximately

$$\frac{4V_o}{\sqrt{n}} = \frac{4x19}{\sqrt{25{,}000}} = 0.481 \text{ ml}$$

Thus, an approximate value for the peak capacity will be given by

$$\frac{\text{Pore Volume}}{\text{Peak Volume}} = \frac{8}{0.481} = 16$$

This value appears to be in reasonable agreement with the peak capacity demonstrated by the chromatogram shown in figure 2. It is seen that the micro-reticulated gel gives separations as good as, if not better than, those obtained on silica gel and all the solutes are adequately separated for quantitative assessment.

A more typical application of micro-reticulate resins in exclusion chromatography is shown by the separation of a standard protein mixture depicted in figure 3.

Figure 3

The Separation of a Standard Protein Mixture by Exclusion Chromatography on a Vinyl Alcohol-Styrene Co-Polymer Hard Gel

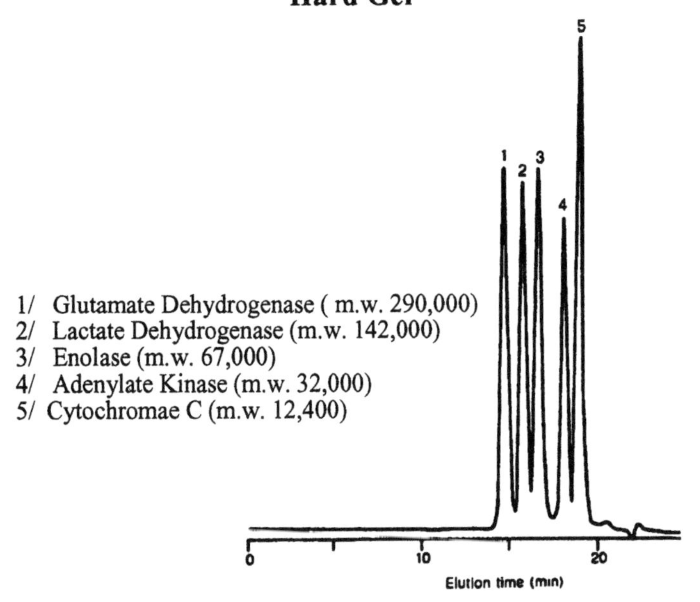

Courtesy of Asahipak Inc.

The column used was 50 cm long and 7.6 mm wide giving a total column volume of about 22.7 ml. This would seem to be in conflict with the value for the dead volume which, from the chromatogram, appears to be about 22 ml (the retention time of 22 min at a flow rate of 1 ml/min). It would appear that the resin occupied no volume in the column. This paradox could be explained on the basis that the separation was not achieved solely by exclusion. The dispersive and polarizable nature of the resin was contributing dispersive and possibly slight polar interactions with the solutes increasing their retention above that expected from exclusion alone. This would be supported by the fact that the mobile phase contained no solvent and consisted of a buffer solution containing 0.1 M sodium phosphate and 0.3 M sodium chloride (pH 7.0) which would have minimum dispersive interactions with the solute. In contrast, the hydrocarbon matrix of the styrene gel would be capable of exerting strong dispersive interactions with the dispersive groups present in the proteins. The use of mixed retention systems to achieve the separation is perfectly acceptable and occurs in most distribution systems. The stationary phase was GFA-50, the number 50 referring to the columns length. The pore size was defined in terms of the minimum molecular weight of a totally excluded solute which was given as 300,000 and the material was specifically prepared for exclusion chromatography.

The particle size was 5 μm and thus the expected efficiency would be in the region of

$$\frac{50}{.001} = 50,000 \text{ theoretical plates}$$

Now from figure 3 it is seen that the retention volume is about 22 ml and thus a mean value for the peak volume can be calculated.

$$\frac{4V_o}{\sqrt{n}} = \frac{4x22}{\sqrt{50,000}} = 0.393 \text{ ml}$$

The peak capacity can be calculated approximately using the corrected retention volume of the last peak instead of the pore volume as there is some interactive retention.

$$\frac{\text{Corrected Retention Volume}}{\text{Peak Volume}} = \frac{22-14}{0.393} = 20$$

It is seen that the peak capacity realized is far less than would be expected from the approximate calculation. This, in fact, is not surprising due to the size of the solute molecules. The diffusivity of the large solute molecules in either phase is so low that the resistance to mass transfer terms become inordinately large. Consequently, when operating significantly above the optimum velocity, very poor efficiencies are obtained.

The analyst should be aware that generally, when working with large molecules (particularly proteins), column efficiencies may, under some circumstances, be as little as 10% of the expected value and this must be taken into account when choosing the column and the phase system.

Chiral Separations

The anxiety that arose from the birth defects that were caused by the drug thalidomide evoked great concern regarding the relative physiological activity of the different geomeric isomers in all pharmaceutical products. As a result, the relative amounts of the different isomers present in virtually all chiral drugs are now required to be assayed and their physiological activity identified. Chiral separations are fundamentally entropically driven. The isomers that fit closest to the chiral sites on the stationary phase will be most restricted in freedom of movement and thus will have the largest retention. Those that fit less closely to the surface will have greater freedom and, consequently, will be the least retained.

The theory of the separation of geometric isomers on stationary phases that have a number of sterogenic centers has not been developed to the point where a particular stationary phase together with an appropriate mobile phase can be deduced for the separation of a specific pair of isomers. A number of theories have been put forward to explain the resolution of geometric isomers (some of which have been quite "imaginative" and "colorful") yet a reliable theory to help in phase selection for a hitherto unresolved chiral pair is still lacking. Unfortunately, the analyst is left with only two alternatives. The first is to search the literature for a model separation similar to the problem in hand and start with that phase system or, alternatively, resort to the technique of the early days of LC, namely, find the best phase system by a trial-and-error routine.

There are two basic procedures that have been successfully used for the separation of isomers. The first is to add a chiral agent to the mobile phase such that it is adsorbed, for example, on the surface of a reverse phase, producing a chirally active surface. This approach has been discussed on page (38) in chapter 2. The alternative is to employ a stationary phase that has been produced with chiral groups bonded to the surface.

The more useful types of chirally active bonded phases are those based on the cyclodextrins. There are a number of different types available, some of which have both dispersive or polar groups bonded close to the chirally active sites to permit mixed interactions to occur. This emphasizes the basic entropic differences between the two isomers being separated. A range of such products is available from ASTEC Inc. and a separation of the *d* and *l* isomers of scopolamine and phenylephrine are shown in figure 4. The separations were carried out on a cyclodextrin bonded phase (CYCLOBOND 1 Ac) that had been acetylated to provide semi-polar interacting groups in close proximity to the chiral centers of the cyclodextrin. The column was 25 cm long, 4.6 mm in diameter and packed with silica based spherical bonded phase particles 5μm in diameter. Most of the columns supplied by ASTEC Inc. have these dimensions and, consequently, provide a limited maximum efficiency of about 25,000 theoretical plates at the optimum velocity. A limited efficiency will, in turn, demand a certain separation ratio between the isomers before resolution will be realized.

Figure 4

The Separation of the Isomers of Scopolamine and Phenylephrine on a Cyclodextrin Bonded Phase

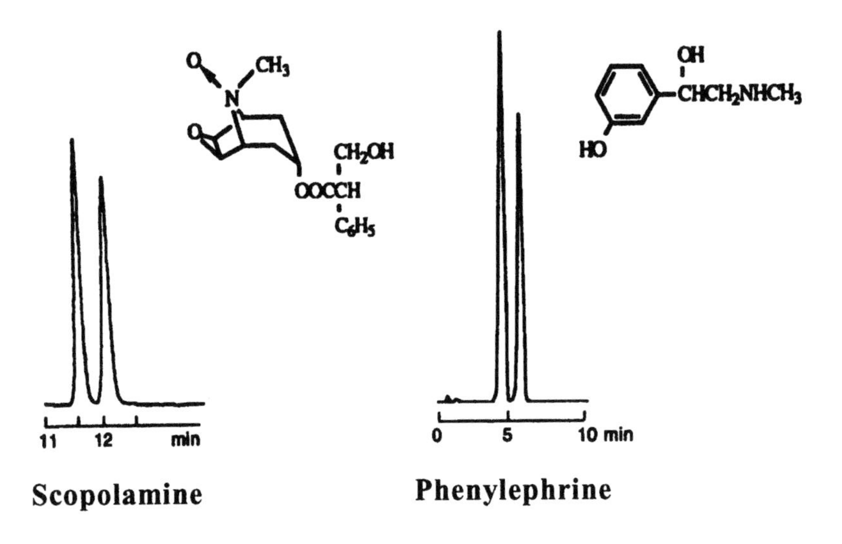

Courtesy of ASTEC Inc.

Re-stating the equation that provides the number of theoretical plates (n) necessary to separate a pair of solute, having a separation ratio of (α), and the first of the pair, (A), eluted at (k'_A)

$$n = \left(\frac{4(1 + k'_A)}{k'_A(\alpha - 1)} \right)^2$$

Rearranging,

$$\alpha = 1 + \frac{4(1 + k'_A)}{k'_A \sqrt{n}}$$

Thus, the minimum value of (α) for any pair of solutes can be calculated for any given column. The minimum values of (α) required for a pair of solutes that will be separated on a column having 25,000 theoretical plates (the efficiency of the standard ASTEC column 25 cm long, 4.6 mm in diameter and packed with spherical particles 5 μ in diameter) is shown plotted against the (k') of the first eluted solute is shown in figure 5.

Figure 5

Graph of Minimum Separation Ratio for a Solute Pair that Can Be Separated on a Column of 25,000 Theoretical Plates against Capacity Ratio of the First Eluted Solute

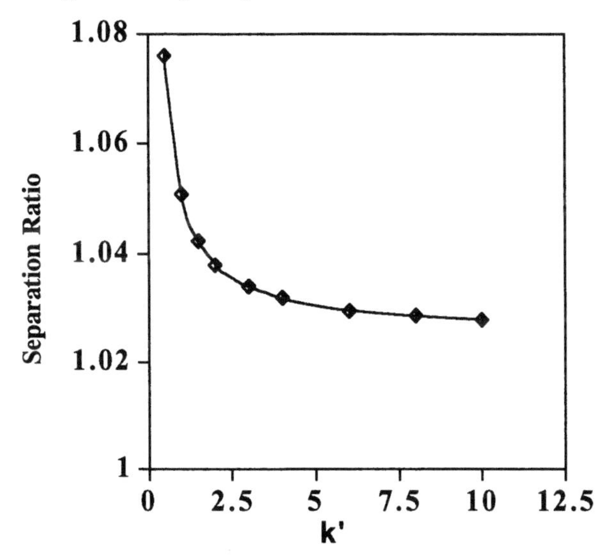

It is seen that, providing the isomers are eluted at a (k') value greater than about 2.0, the column will separate those solute pairs having separation ratios as low as about 1.03. This, however, assumes that the column is very well packed and is operated at about the optimum mobile phase velocity. In practice, a more realistic minimum separation ratio would be between 1.035 and 1.04. However, as it will be seen, the cyclodextrin bonded phases can easily provide separation ratios significantly greater than these values. This is achieved by introducing appropriate interacting groups proximate to the chiral sites to augment the entropic differential retention by additional polar and dispersive interactions. The scopolamine was separated employing a mobile phase consisting of a 2% solution of acetonitrile in a 0.5% aqueous solution of triethylamine buffered to a pH of 4.1 with glacial acetic acid. The two isomers were eluted at 8.7 and 10.9 minutes respectively which, assuming the dead time to be relatively small, gives a separation ratio of about 1.25. The phenylephrine was separated using a 10% methanol with 90% of a 0.5% aqueous sodium acetate solution buffered to a pH of 6.0. The two isomers were eluted at 6.97 min and 7.73 min respectively giving a separation ratio of approximately 1.109. It is clear the column efficiencies were more than adequate to achieve complete resolution and, consequently, accurate quantitative results would be obtainable.

Modern bonded phases of the Cyclobond type can achieve large separation ratios between the different isomers using mixed retention mechanisms. By bonding appropriate dispersive or polar groups close to the sterogenic centers of the cyclodextrin, the retention, entropically driven by geometric fitting, is augmented by polar and/or dispersive interactions with neighboring groups. The isomer that is geometrically isolated from the stereogenic site cannot interact with the neighboring groups and, consequently, the differential interaction of the isomers is amplified.

The mixed retention mechanism described above has a parallel in the effect of exclusion on retention when using porous stationary phases. The smaller molecules can enter more pores and thus interact with more stationary phase than the larger molecules that are excluded from many pores and, consequently, interact with less stationary phase. Assuming the smaller molecules are more strongly retained, the exclusion of large molecules augments the difference in retention between molecules of different size.

A further example of mixed retention mechanisms on the resolution of geometric isomers is afforded by the separation shown in figure 6.

Figure 6

The Separation of the Isomers of Methadone, Norphenylephrine and t-boc-Alanine by Mixed Retention Mechanisms

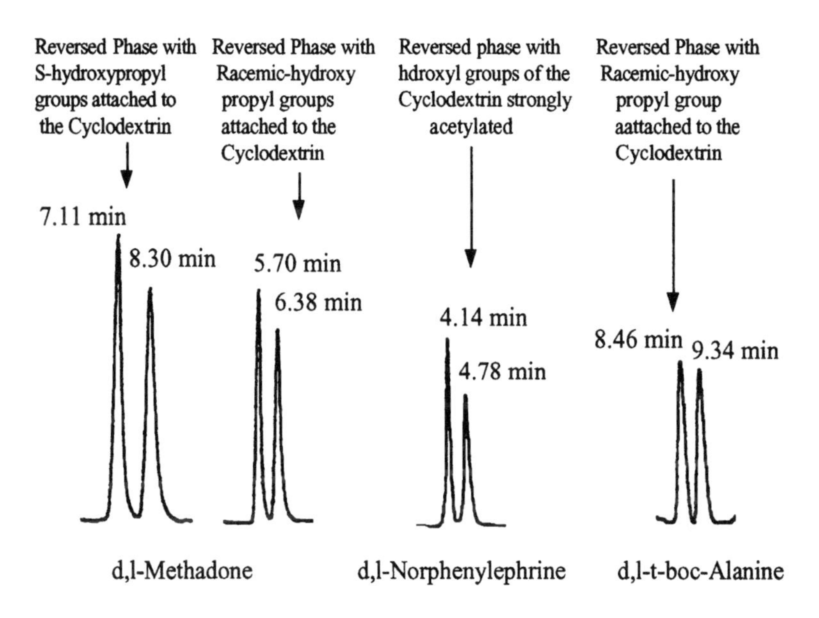

It is seen that methadone is shown separated on two stationary phases, one a bonded cyclodextrin carrying the S-hydroxypropyl groups and the other a cyclodextrin carrying the racemic hydroxypropyl groups. The two isomers of methadone were eluted at 7.11 and 8.30 minutes respectively from the phase with the S-hydroxypropyl group. In contrast, the two methadone isomers were eluted at 5.70 and 6.38 minutes, respectively, from the material carrying the racemic hydroxypropyl groups. The separation ratio of the isomers was 1.119 from the phase with the racemic hydroxypropyl groups and 1.167 from the phase with the 'active' hydroxypropyl groups. Thus, although the absolute retention was higher from the bonded phase with the racemic material, it is not clear whether this was due to a different retentive mechanism, flow rate or bonded phase loading. The interesting point is, however, that the S-hydroxypropyl group only increases the separation ratio by about 4% and thus, it is the stereogenic centers of the cyclodextrin that are still playing the major role in the chiral selectivity. It is also seen that very high separation ratios are obtained for the isomers of *d,l* norphenylephrine (1.155) and *d,l* t-boc-alanine (1.104). Such high differential retention now obtainable between the optically active components of complex mixtures has simplified chiral separations considerably.

The Cyclobond materials are some of the most effective in separating isiomers generally and their development continues. It is likely that chiral chromatography will become increasingly important as the products from biotechnology continue to proliferate into the pharmaceutical field.

Interactive LC Systems

Interactive LC systems are those where solute retention is predominantly controlled by the relative strengths of the molecular interactions between solute molecules with those of the two phases. In such systems, exclusion and entropically driven interactions will be minor contributions to retention. The three basically different types of molecular interaction, dispersive, polar and ionic give rise to three subgroups, each subgroup representing a separation where one specific type of interaction dominates in the stationary phase and thus governs solute retention. The subgroups are as follows:

1/ Dispersive Interaction Chromatography
2/ Polar Interaction Chromatography
3/ Ionic Interaction Chromatography
4/ Composite Interaction Chromatography

The fourth group, Composite Interaction Chromatography, involves more than one type of interaction and no one type of interaction dominates.

Dispersive Interaction Chromatography

Dispersive interaction chromatography is carried out using the well-known reverse phases, usually silica gel based, but dispersive phases based on polystyrene divinyl benzene resins are know becoming popular. There are three classes of reverse phase generally available and they are the C4, C8, and the C18 reverse phases, the numbers referring to the number of carbon atom in the reverse phase chain. The intensity of the dispersive interactions between the solutes and the reverse phase increases with the chain length and thus, the C4 chain exhibits the weakest interaction. This reverse phase, however, is often used when separating proteins, as the stronger interaction between the proteins and the phases with longer chain lengths can be sufficiently severe to cause denaturation. It should be noted, however, that when separating proteins for analytical purposes, the peaks are not usually collected for other work, and so it does not matter if the protein is denatured.

The C18 reverse phase exhibits the maximum dispersive interactions with the solutes and is thus, chosen when the difference in dispersive character of the solutes is small or subtle. Employing a C18 reverse phase accentuates the dispersive interactions with the solutes and consequently improves their relative retention. C18 columns also exhibit a somewhat higher loading capacity and so large charges can be placed on the column before overload occurs. This can be useful in trace analysis, where large charges are often necessary to detect the minor components at a level where they can quantitatively evaluated.

The C8 (octyl reverse phase) is the general "work horse" of reverse phases and is recommended as the first to be tried when attempting to exploit dispersive interaction to achieve a separation. Columns packed with C8 material are available in range of lengths from 3 to 50 cm long and can be packed with particles 3, 5, 10 and 18 m in diameter. Consequently, a wide range of column efficiencies is available to the analyst, which, in the methods development laboratory, should always be readily accessible.

The mobile phases that are most effective for use with reverse phases are aqueous mixtures of methanol or acetonitrile and for subtle adjustments, ternary mixtures of water, methanol and acetonitrile or tetrahydrofuran can be used. The greater the water content the more the solutes with dispersive groups will be retained and in fact, in pure water, many substances are irreversibly held on a reverse phase. As already discussed, this characteristic make reverse phases very useful for solute extraction and concentration from aqueous solutions prior to analysis.

Methanol/water is the most polar solvent mixture and the least dispersive, so methanol/water mixtures will offer largely competing polar interactions to the solutes against the dispersive interaction between the solutes and the reverse phase. Acetonitrile, however, is more dispersive than methanol and less polar, so aqueous solutions of acetonitrile offer greater competing dispersive interactions than methanol against those of the stationary phase. Tetrahydrofuran is even more dispersive than acetonitrile and is rarely used alone with water but as a small component added to a methanol/water, or to acetonitrile/water mobile phase. The effect of the tetrahydrofuran to increase the dispersive character of the mobile phase without changing the polar characteristics significantly. The solvent system can be adjusted in a very subtle manner to achieve the desired selectivity, but the rationale for choosing the mixture must be based on the interactive character of the solutes and the two phases.

An example of the separation of a mixture of explosives on a C8 column is shown in figure 7. The column was 3.3 cm long, 4.6 mm in diameter and packed with 3 μm C8 silica based reverse phase. This short column has a potential efficiency of 5,500 theoretical plates.

The flow rate was 2 ml/min and as the last peak was eluted about 3.2 min the retention volume was 6.4 ml.

Thus, $$\frac{4V_o}{\sqrt{n}} = \frac{4x6.4}{\sqrt{5,500}} = 0.345 \text{ ml}$$

Figure 7

The Separation of a Mixture of Explosives on a C8 Reverse Phase Column

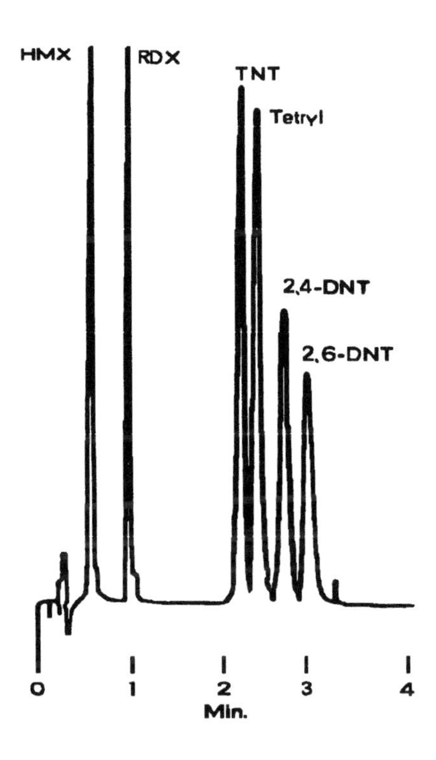

Courtesy of Supelco Inc.

Now the dead time appeared to be about 0.6 min, which is equivalent to about 1.2 ml, consequently the peak capacity is approximately

$$\frac{\text{Corrected Retention Volume}}{\text{Peak Volume}} = \frac{6.4 - 1.2}{0.345} = 15$$

It would appear that a peak capacity close to 15 is probably realized and the separation is quite adequate for accurate quantitative assessment.

The mobile phase used was 2% tetrahydrofuran (THF) in a mixture of 30% methanol and 70% water . This is an interesting example of the use of a small quantity of THF to increase the dispersive character of the mobile phase while maintaining the high polarity of the methanol water mixture. To achieve the same increase in dispersive interactions by increasing the methanol content would probably require as much as 40-45% methanol. At this concentration the polarity of the mobile phase would have drastically changed and the selectivity of the system for the more polar materials probably lost. It is also seen that the overall sensitivity of the system is high, components being present at a level of about 100 ng.

Another example of the use of a C8 column for the separation of some benzodiazepines is shown in figure 8. The column used was 25 cm long, 4.6 mm in diameter packed with silica based, C8 reverse phase packing particle size 5 μ. The mobile phase consisted of 26.5% v/v of methanol, 16.5%v/v acetonitrile and 57.05v/v of 0.1M ammonium acetate adjusted to a pH of 6.0 with glacial acetic acid and the flow-rate was 2 ml/min. The approximate column efficiency available at the optimum velocity would be about 15,000 theoretical plates. The retention time of the last peak is about 12 minutes giving a retention volume of 24 ml.

Thus, the peak volume is thus given by

$$\frac{4V_0}{\sqrt{n}} = \frac{4x24}{\sqrt{15,000}} = 0.784 \text{ ml}$$

The dead time is difficult to estimate but appears to be about 0.75 minutes which is equivalent to a dead volume of about 1.5 ml. Consequently, the peak capacity will be given by

$$\frac{\text{Corrected Retention Volume}}{\text{Peak Volume}} = \frac{24-1.5}{0.784} = 28.7$$

It is clear that the chromatogram in figure 8 does not have a peak capacity of 29 although a perfectly satisfactory resolution is obtained.

Figure 8

The Separation of Eight Diazepines Employing a C8 Reverse Phase Column

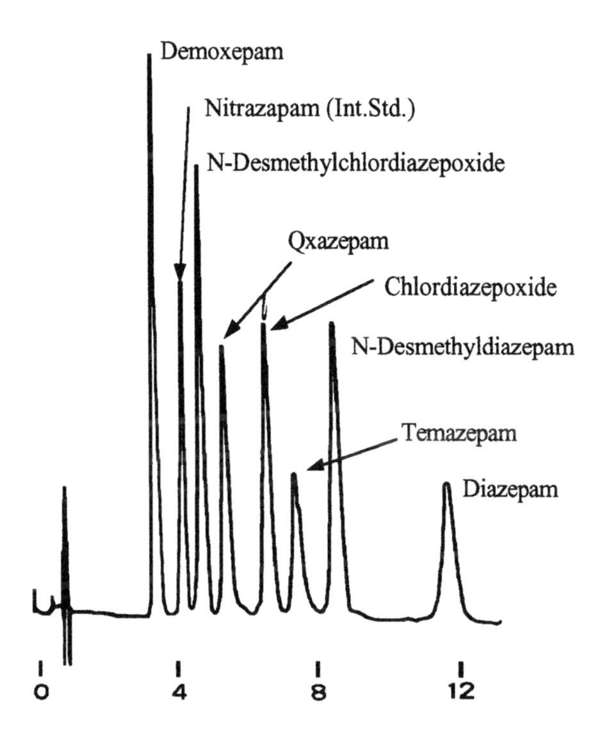

Courtesy of Supelco Inc.

It should be noted that at a flow rate of 2 ml/min., the mobile phase velocity will be well above that of the optimum and so the maximum efficiency will not be realized. Generally, when there are more theoretical plates available than required, the flow rate is increased until the separation required is just realized. This procedure trades efficiency for time and allows the separation to be achieved in the minimum time *given the column and phase system that has been chosen*. It must be emphasized that the minimum analysis time can only be achieved using the optimum column which will be unique for the sample and phase system chosen. The design of optimum columns, however, is outside the scope of this book.

The mobile phase is interesting in that the water is buffered appropriately to complement the dissociation constants of the solutes. A mixture of methanol and acetonitrile is employed, the acetonitrile being used to increase the dispersive interactions in the mobile phase. The reason for the particular solvent mixture is not clear and it would appear that the separation might be achieved equally well by using a stronger solution of methanol alone or a more dilute solution acetonitrile alone. There is no particular advantage to one solvent mixture over another except for the fact that 'waste' acetonitrile produces greater solvent disposal problems than methanol.

Finally an example is included of the use of a C18 reverse phase column. The packing is also a silica based but is contained in a short column 3.3 cm long, 4.6 mm in diameter and packed with particles 3 µm in diameter. The example of its use is in the separation of mixture of growth regulators which is shown in figure 9.

The peak capacity is not pertinent as the separation was developed by a solvent program. The expected efficiency of the column when operated at the optimum velocity would be about 5,500 theoretical plates. This is not a particularly high efficiency and so the separation depended heavily on the phases selected and the gradient employed. The separation was achieved by a complex mixture of ionic and dispersive interactions between the solutes and the stationary phase and ionic, polar and dispersive forces between the solutes and the mobile phase. The initial solvent was a 1% acetic acid and 1 mM tetrabutyl ammonium phosphate buffered to a pH of 2.8. Initially the tetrabutyl ammonium salt would be adsorbed strongly on the reverse phase and thus acted as an adsorbed ion exchanger. During the program, acetonitrile was added to the solvent and initially this increased the dispersive interactions between the solute and the mobile phase.

Figure 9

The Separation of a Mixture of Growth Regulators on a C18 Reverse Phase Column

1/ 6-Benzylaminopurine riboside (445ng)
2/ Indole-3-acetic acid (222ng)
3/ Ascisic acid (222ng)
4/ α-Naphthalene acetamide (222ng)
5/ Colchicine (555ng)
6/ Indole-3-propanoic acid (333ng)
7/ p-chlorophenoxy acetic acid (333ng)
8/ Indole-3-butyric acid (333ng)
9/ α-Naphthalene acetic acid (333ng)
10/ β-Naphthalene acetic acid (333ng)
11/ 2,4-Dichlorophenoxy acetic acid (555ng)
12/ Indole-3-acetic acid ethyl ester (445ng)
13/ 2,4,5-Trichlorophenoxy acetic acid (555ng)

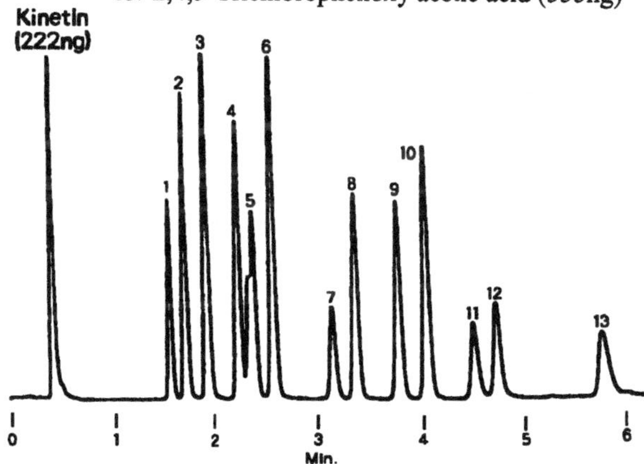

Courtesy of Supelco Inc.

As the acetonitrile concentration increased, however, the concentration of adsorbed tetrabutyl ammonium salt would also be reduced and it would be desorbed from the reverse phase with a resulting reduction in the ionic interactions of the solutes with the stationary phase. At higher concentrations of acetonitrile, the tetrabutyl ammonium salt would be completely desorbed from the stationary phase and the interactions of the solutes with the stationary phase would become almost exclusively dispersive. This is an example where the phase system is complex and is required to be so, because limited column efficiency was available. One wonders if a column with intrinsically more theoretical plates might have achieved the separation with a simpler solvent system and a more straightforward solvent program.

Polar Interaction Chromatography

Although, by employing reverse phases, it is possible to have interactions with the stationary phase that are exclusively *dispersive*, it is not possible to have exclusively *polar* interactions with a stationary phase. This is because all compounds, or groups, that are polarizable or have permanent dipoles, also exhibit dispersive characteristics. The hydroxyl group probably has the minimum capacity for dispersive interactions and so silica gel, with its interacting surface hydroxyl groups, will come close to a stationary phase that provides solely polar interactions. However, it must be stressed that even with silica gel, some dispersive interactions between the solute molecules and the surface will occur. In contrast, the so called 'polar' bonded phase, whether carrying -CN groups or -OH groups or other polar moieties, will also be associated with some alkyl side chains. Consequently, there will be strong dispersive interactions between the side chains and the solute molecules as well as polar interactions with the polar group or groups.

The silica gel surface is extremely polar and, as a result, must often be deactivated with a polar solvent such as ethyl acetate, propanol or even methanol. The bulk solvent is usually an n-alkane such as n-heptane and the moderators (the name given to the deactivating agents) are usually added at concentrations ranging from 0.5 to 5% v/v. Silica gel is very effective for separating polarizable materials such as the aromatic hydrocarbons, nitro hydrocarbons (aliphatic and aromatic), aliphatic ethers, aromatic esters, etc. When separating polarizable substances as opposed to substances with permanent dipoles, mixtures of an aliphatic hydrocarbon with a chlorinated hydrocarbon such as chlorobutane or methylene dichloride are often used as the mobile phase. Such solvents ensure that only dispersive interactions take place in the mobile phase whereas the polar interactions are confined to silica gel stationary phase. As a result both the retention and the selectivity of the stationary phase will be based on polarity differences of the solutes. After a separation using a specific solvent with a moderator, the silica may need to be changed to a different activity for the next sample. Bringing the silica back to a new equilibrium with another solvent and moderator may take some time and a considerable amount of solvent. This is one of the disadvantages of native silica when used as an LC stationary phase. Bonded phases rapidly come into equilibrium after a solvent change which makes them easier to use with gradient elution and is one of the factors that accounts for their popularity.

An example of a separation primarily based on polar interactions using silica gel as the stationary phase is shown in figure 10. The macro-cyclic tricothecane derivatives are secondary metabolites of the soil fungi *Myrothecium Verrucaia.* They exhibit antibiotic, antifungal and cytostatic activity and, consequently, their analysis is of interest to the pharmaceutical industry. The column used was 25 cm long, 4.6 mm in diameter and packed with silica gel particles 5 μ in diameter which should give approximately 25,000 theoretical plates if operated at the optimum velocity. The flow rate was 1.5 ml/min, and as the retention time of the last peak was about 40 minutes, the retention volume of the last peak would be about 60 ml.

Using the usual equation, the peak volume is given by

$$\frac{4V_o}{\sqrt{n}} = \frac{4x60}{\sqrt{25,000}} = 1.517 \text{ ml}$$

The dead volume is difficult to estimate, but for a column of that size would be about 2.5 ml and, consequently, the peak capacity will be

$$\frac{\text{Corrected Retention Volume}}{\text{Peak Volume}} = \frac{60-2.5}{1.517} = 38$$

It is seen that a high peak capacity was available but, as a gradient program was used, the isocratic peak capacity is not pertinent. The mobile phase program started with a solvent mixture containing 20% v/v of ethyl acetate in n-hexane and ended with pure ethyl acetate.

Figure 10

The Separation of Some Tricothescenes on Silica Gel

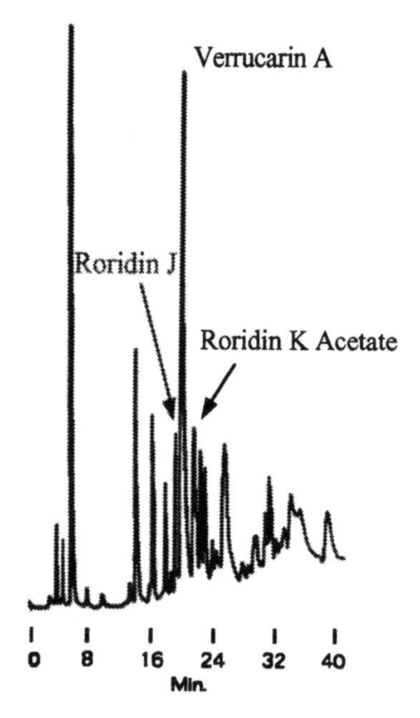

Courtesy of Supelco Inc.
(Supplied to Supelco by Dr. B. B. Jarvis, University of Maryland)

Another interesting example of the use of native silica (shown in figure 11) is for the analysis of Darvocet® and its generic equivalent formulation. Darvocet® is an acetaminophen product and the active ingredient and other substances present are weakly polar and thus lend themselves to separation on silica gel. It is seen that the analysis was completed in less than 4 minutes using a short column 3.3 cm long and 4.6 mm in diameter. The silica packing had a particle size of 3 μ providing a maximum efficiency of about 5,500 theoretical plates.

Figure 11

The Analysis of Acetaminophen Formulations

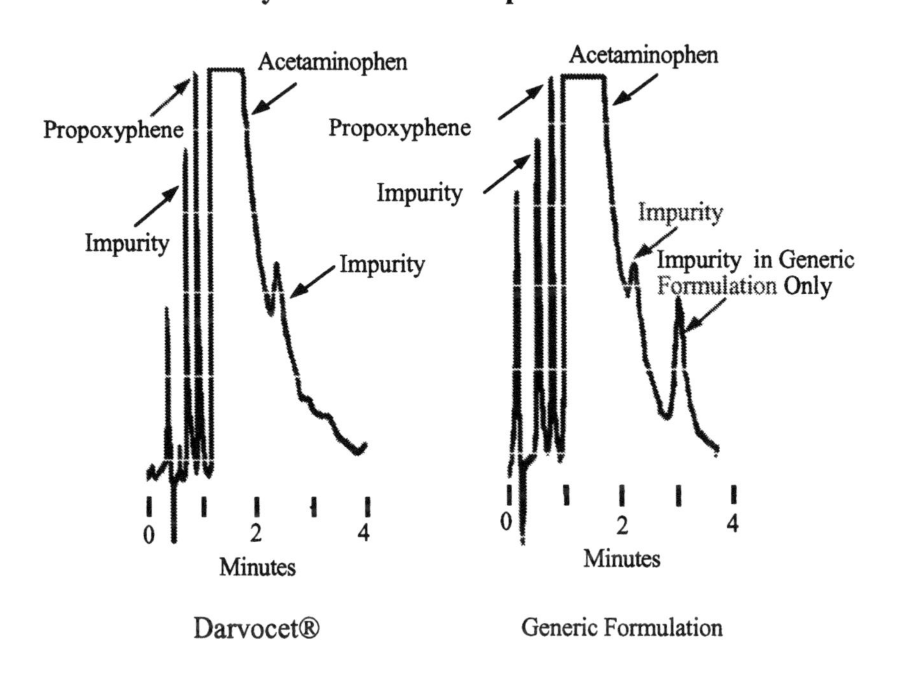

Courtesy of Supelco Inc.

It is seen that to identify the impurities, the column appeared to be significantly overloaded. Nevertheless, the impurities were well separated from the main component and the presence of a substance was demonstrated in the generic formulation that was not present in the Darvocet®. The mobile phase was 98.5% dichloromethane with 1.5% v/v of methanol containing 3.3% ammonium hydroxide. The ammoniacal methanol deactivated the silica gel but the interaction of the solutes with the stationary phase would still be polar in nature. In contrast solute interactions with the methylene dichloride would be exclusively dispersive.

A final example of the use of silica gel as an interactive stationary phase is afforded by the separation of some steroid hormones shown in figure 12.

Figure 12

The Separation of Steroid Hormones

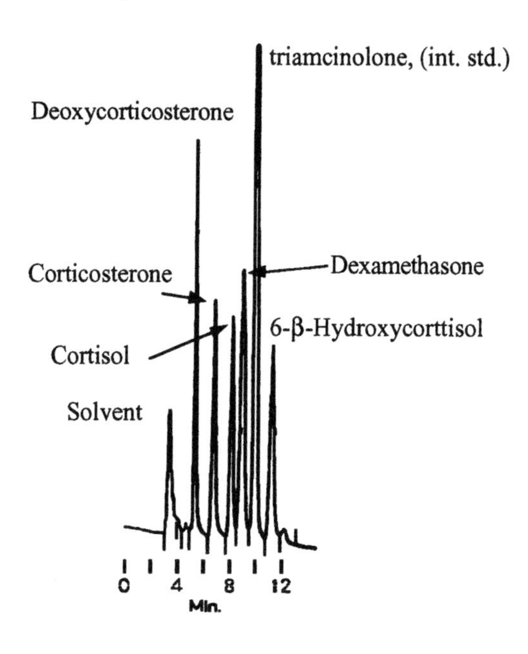

Courtesy of Supelco Inc.
(Supplied to Supelco by Dr. S. N. Rao and Prof. M. Okamato, Cornell University Medical College, New York)

The column used was 25 cm long, 4.6 mm in diameter, and packed with silica gel particle (diameter 5 μm) giving an maximum efficiency at the optimum velocity of 25,000 theoretical plates. The mobile phase consisted of 76% v/v n-hexane and 24% v/v 2-propyl alcohol at a flow-rate of 1.0 ml/min. The steroid hormones are mostly weakly polar and thus, on silica gel, will be separated primarily on a basis of polarity. The silica, however, was heavily deactivated by a relatively high concentration of the moderator 2-propyl alcohol and thus the interacting surface would be covered with isopropanol molecules. Whether the interaction is by sorption or displacement is difficult to predict. It is likely that the early peaks interacted by sorption and the late peaks by possibly by displacement.

Ionic Interaction Chromatography

Ionic interaction chromatography, or ion chromatography as it is usually called, is typically carried out employing ion exchange resins as the stationary phase. There are some silica based ion exchange materials available, but due to the instability of bonded silicas to high salt concentrations and extremes of pH, they have very limited areas of application. The polystyrene divinyl benzene cross-linked polymers, on the other hand are completely stable to a wide range of salt concentrations and can function well within the pH range of 2.0 to 12.0. The obvious use of ion exchange chromatography is in the separation of all types of anions and cations and a number of examples of such applications have already been given. Metal cations and inorganic anions are all separated predominantly by ionic interactions with the ion exchange resin. Organic acids and bases, however, will be retained by mixed interactions with the stationary phase, as dispersive and polar interactions will take place between the solute molecules and the aromatic nuclei and the aliphatic side chains of the base polymer. In the separation of simple acids and bases it is important that the mobile phase is buffered appropriately according to the pK_a of the salts so that dissociation occurs and the ions are free to interact with the stationary phase. By employing mobile phase additives and using novel operating conditions, ionic interactions can be used to separate a far wider range of materials than simple organic and inorganic anions and cations.

An example of the use of ion interactions to separate the non-selective contact herbicides Diquat and Paraquat is given in figure 13. The column was fairly large (10 cm long and 7.6 mm in diameter) and was packed with cross-linked polystyrene beads, 9 μm in diameter, carrying -COOH groups as the interacting ion moieties. At the optimum flow rate the column would give about 5,500 theoretical plates but at the flow rate used which was 1 ml/min the efficiency would be considerably below that. Nevertheless, a very good separation was obtained.

Figure 13

The Separation of the Herbicides Diquat and Paraquat

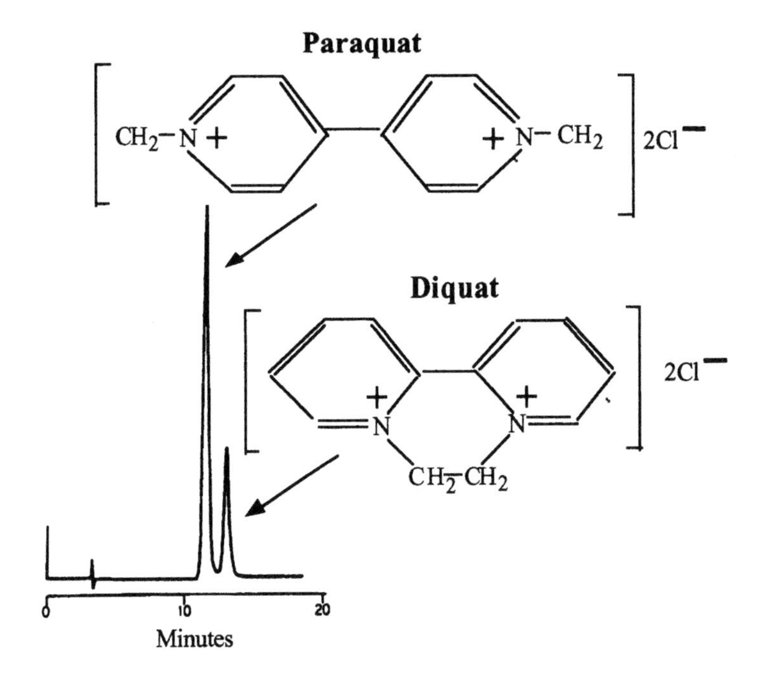

Courtesy of Asahi Chemical Industry Co. Ltd.

The mobile phase was an aqueous solution containing 50 mM sodium phosphate and 150 mM sodium chloride at a pH 7.0. Although ionic interactions are likely to constitute the major contribution to retention and selectivity, there would also be significant polar interactions and some dispersive interactions between the aromatic nuclei of the solutes and the aromatic nuclei and the aliphatic side chains of resin respectively. Under these circumstances, without considerable experimental work, it is impossible to identify the relative magnitude of the different contributions from each type of interaction.

As a result of the high activity in the biotechnology field, one of the most popular and important applications of ion exchange chromatography is for the separation of proteins. In general, proteins do not have strong ionic moieties with which ion exchange groups can interact in fact, the basic separation of proteins by ion exchange interactions is somewhat of an anomaly. Other types of interaction between the protein and the resin will, without doubt, play a part but they do not appear to be dominant in the retentive mechanism. Proteins can be eluted by increasing salt concentration in the mobile phase which will have little effect on any polar or dispersive interactions the proteins may have with the resin. An example of a protein separation on an ion exchange resin is shown in figure 14.

Figure 14

The Separation of a Protein Mixture by Ion Exchange Resin

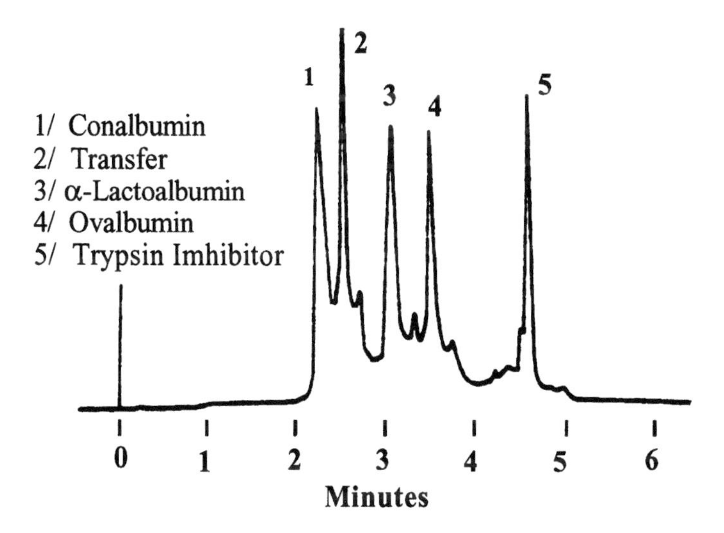

Courtesy of TOYO SODA Manufacturing Co., Ltd.

The column, designated as TSKgel DEAE-NPR a weak anion exchanger, was 3.5 cm long and 4.6 mm in diameter packed with non-porous resin beads 2.5 μ in diameter. Thus, the maximum efficiency available at the optimum mobile phase velocity would be about 7,000 theoretical plates. The sample was a crude hexokinase product and an excellent separation was obtained. It should be noted that the peaks, despite being for high molecular weight proteins, have been constrained to reasonable widths. The flow rate was 1.5 ml/min (well above the optimum flow rate) and a linear salt gradient was used to develop the separation starting at 0 and ending at 0.5 M sodium chloride solution in 20 mM tris-HCl buffer (pH 8.0). It is claimed that non-porous ion exchange resins give better separations than the porous resins, this improvement is achieved, however, at the expense of a relatively low loading capacity. Stationary phases with a low loading capacity are not, generally, a problem in analytical ion exchange chromatography.

Ion chromatography can be used in unique ways and by appropriate modification can often be applied to the separation of mixtures where the components themselves do not ionize or do not normally produce interactive ions in aqueous solution. A good example of this type of separation is afforded by the analysis of saccharide mixtures using ion exchange interactions. An illustration of such a separation is given in figure 15.

The separation is achieved by reacting the saccharides with a borate which readily forms complex anions. The complex is simply formed by including a borate buffer in the mobile phase. The process is, in fact, reminiscent of 'in-line' derivatization. The packing, designated as a TSKgel Sugar AXG was a strong anion exchange resin, had a particle diameter of 10 μ and contained quaternary ammonium ions as the exchange moiety. The material was packed in a column 15 cm long and 4.6 mm in diameter and consequently would have a potential efficiency of 7,500 theoretical plates. The mobile phase consisted of three borate buffer solutions that were employed in a stepwise gradient. The first was a 0.5 M borate buffer (pH 7.7), the second a 0.7 M borate buffer (pH 7.3) and the last a 0.7 M borate buffer (pH 8.7). The flow rate employed was 4 ml/min so an efficiency fairly close to the maximum would have been obtained, providing the column was packed well.

Figure 15

The Separation of a Saccharide Mixture by Ion Exchange Chromatography

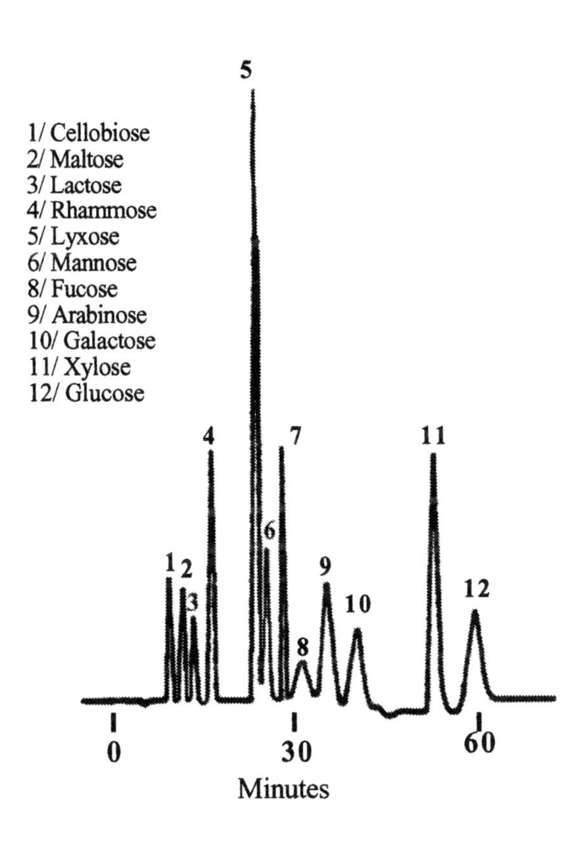

Courtesy of TOYO SODA Manufacturing Co., Ltd.

The last example clearly introduces an entirely different approach to the separation of a mixture. It is seen that it is just as feasible for the solutes to be modified to suit a particular phase system as it is to choose or modify a phase system to suit the solutes. One of the delights in the practice of analytical chromatography is the wide range of variables and alternative approaches from which the analyst can choose to handle a particular sample.

Mixed Interaction Chromatography

Virtually all interactive mechanisms that control retention in chromatography are, in fact, mixed interactions as shown by the previous application examples. It has already been suggested that reverse phases can exhibit almost exclusively dispersive interactions with solutes. However, as they are almost always employed with aqueous solvent mixtures then, polar and dispersive interactions will still be operative in the mobile phase. Consequently, the examples given here will be taken where the mixed interactions are either unique or represent a separation of special interest.

Figure 16

The Separation of a Ferredoxin Mixture

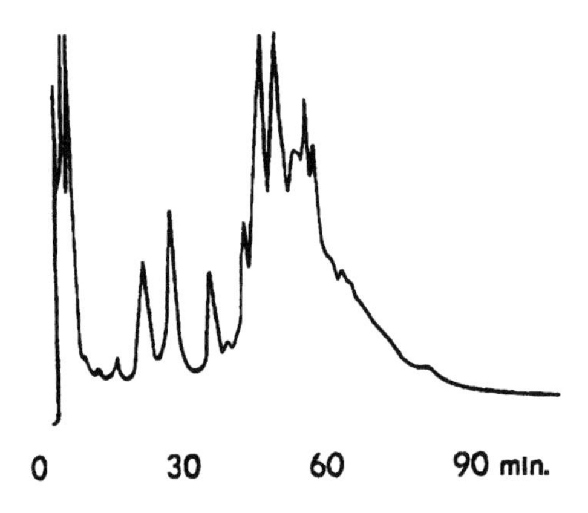

Courtesy of the TOYO SODA Manufacturing Co. Ltd.

The first example will be the separation of a ferredoxin mixture using a bonded phase that contains aromatic nuclei as well as aliphatic chains. The stationary phase will thus, exhibit polar interaction from induced dipoles if the aromatic ring comes into contact with a strong dipoles of the solute and, at the same time, exhibit dispersive interactions between the aliphatic chains and any dispersive centers of the solute molecule. An example of the separation obtained is shown in figure 16.

The ferredoxins are a group of electron transfer factors found in plants and bacteria which are non-heme iron sulfur proteins and which are important agents in photosynthesis, nitrogen and carbon dioxide fixation and respiration. The crude extracts are often difficult to resolve into their components and the separation shown in figure 16 was achieved using a TSkgel Phenyl-5PW column, the stationary phase having phenyl groups exposed on the surface together with aliphatic chains.

The column was 7.5 cm long and 7.5 mm in diameter and packed with particles 10 μ in diameter. Such a column should provide the theoretical maximum efficiency of about 3,750 theoretical plates. However, due to the size of the solute molecules, this efficiency is not likely to be realized which appears to be confirmed by the size of the peaks shown in the chromatogram. It is seen that a flow rate of only 0.5 ml/min was used which, considering the relatively large column diameter, would provide a mobile phase velocity close to the optimum and thus achieve the maximum efficiency possible. The use of the optimum mobile phase velocity also resulted in the very long elution time of 90 minutes. The separation was developed employing a linear gradient extending over 60 min starting with a 0.1 M phosphate buffer (pH 7.0) plus 1.5 M ammonium sulfate and ending with a simple 0.1 M phosphate buffer still at a pH of 7.0. The separation reveals a large number of components present in the mixture and strongly suggests that the separation should be developed further and improved.

Another interesting stationary phase is one that is silica based but contains glyceryl-propyl groups bonded to the surface. The hydroxyl groups of the glycerol provides strong polar interactions with the solute while the aliphatic propyl groups provide centers for dispersive interactions. The separation of some proteins and peptides is shown in figure 17. The separation was achieved fairly rapidly in about 8 minutes and it is interesting to note that the smallest molecule, the di-peptide, is eluted last and thus experiences the greater interaction with the stationary phase.

Figure 17

The Separation of Some Proteins and a Peptide on a Glyceryl-Propyl Bonded Phase

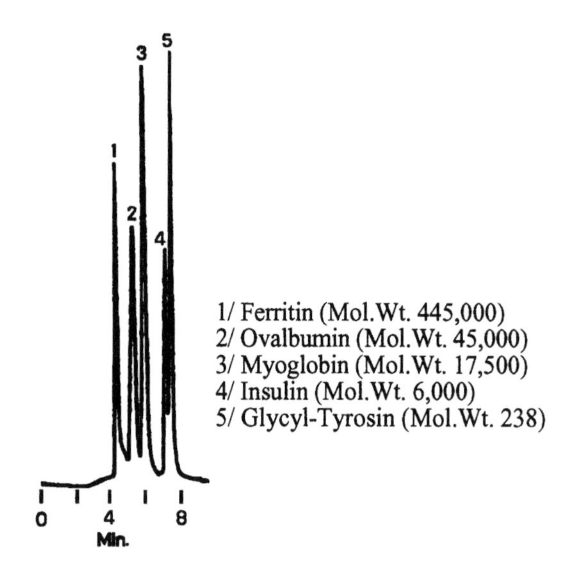

Courtesy of Supelco Inc.

The first solute, ferritin, is an iron storage protein found mainly in the spleen, liver and intestinal mucosa of vertebrates. The column used was 25 cm long, 6.2 mm in diameter and packed with particles 5 μ in diameter. As a consequence, the column should provide a theoretical maximum efficiency of about 25,000 theoretical plates. It is clear that high efficiencies are indeed necessary to separate the materials and that, as a result of the high molecular weight of the early eluted compounds the expected high efficiencies are not realized. This is despite the use of a flow rate of 0.7 ml/min which for a column having a diameter of 6.2 mm would be approaching the optimum value for minimum HETP. Unfortunately, this is inevitable due to the low diffusion rates of large molecules and is not due to any deficiency in the column. The mobile phase was a solution of 0.1 M potassium di-hydrogen phosphate and 0.1 M potassium hydrogen phosphate (pH 6.8). Thus, in the absence of solvent, there were virtually no dispersive forces active in the mobile phase and only polar and ionic forces were used to elute the solutes.

Another type of mixed interaction is the combination of dispersive and polar interactions with the entropic selectivity offered by chiral activity. An example of this is the immobilized isiomer, a 3,5-dinitrobenzoyl derivative of (R)-or (S)-phenylglycine. The electron accepting character of the nitro groups polarizes the aromatic ring to form a strong dipole. Thus, the aromatic ring can exhibit strong polar interactions with any solute isomer that is of the form that can fit closely to the bonded chiral structure.

Figure 18

The Separation of the 2,2,2-trifluoro-1-(9-anthryl) Ethanol Isomers by Mixed Interactions

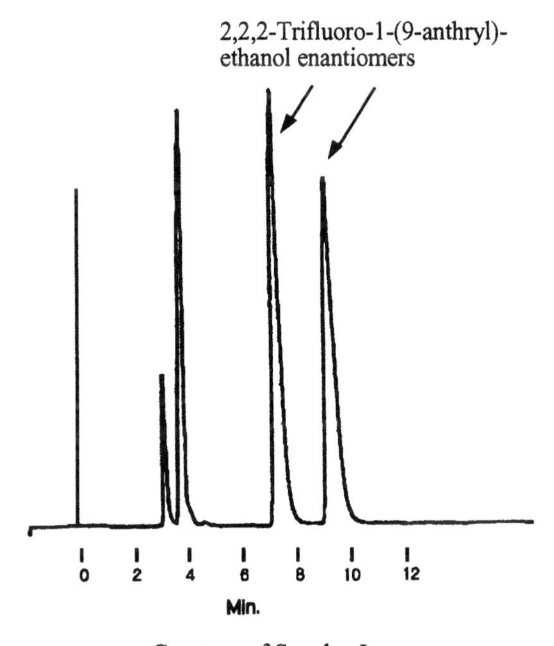

Courtesy of Supelco Inc.

An example of the use of the stationary phase to separate the 2,2,2-trifluoro-1-(9-anthryl) ethanol isomers is shown in figure 18. The column was 25 cm long, 4.6 mm in diameter packed with a silica based bonded phase with a pore size of 100 Å and a particle diameter of 5 μ. The maximum available efficiency would be 25,000 theoretical plates and the mobile phase consisted of a mixture of 5% of isopropyl alcohol and 95% of n-hexane. It is seen that an excellent separation of the two isomers was obtained and the mixed interaction technique has worked very well.

Finally, a somewhat different interpretation of mixed interactions is given in the example of the effect of temperature on column resolution. This subject has been touched upon earlier in the book and it may be recalled that the effect of temperature on the separation process is twofold. Firstly, changing the temperature will change the free energy of distribution and thus retention. In gas chromatography the effect of temperature on retention is profound but the effect is much less in LC systems. This is because the change in free energy with temperature is similar in both phases and thus, the differential change is small. Conversely, the effect of temperature on column efficiency in GC is much less profound than in LC and temperature can dramatically change the width of a solute band eluted from an LC column.

The effect of temperature on the resolution obtained from an LC column is demonstrated unambiguously in figure 19. The mixture consists of a series of monosaccharides and includes one disaccharide and one trisaccharide. The column was a SUPERCOGEL C-611 ion exchange resin contained in a column 30 cm long and 7.8 mm in diameter. It is seen that the effect of temperature on the separation is quite startling. It is seen that raising the temperature *reduces* the retention of the solutes and the separation ratios between them. The change is not great, but is in a direction that would normally *reduce* the separation that was obtained. The efficiency, on the other hand, has increased enormously, so much so, that the reduction in peak width has not only compensated for the reduced separation ratios, but improved the separation to the point of base-line resolution.

Figure 19

The Effect of Column Temperature on the Separation of a Carbohydrate Mixture

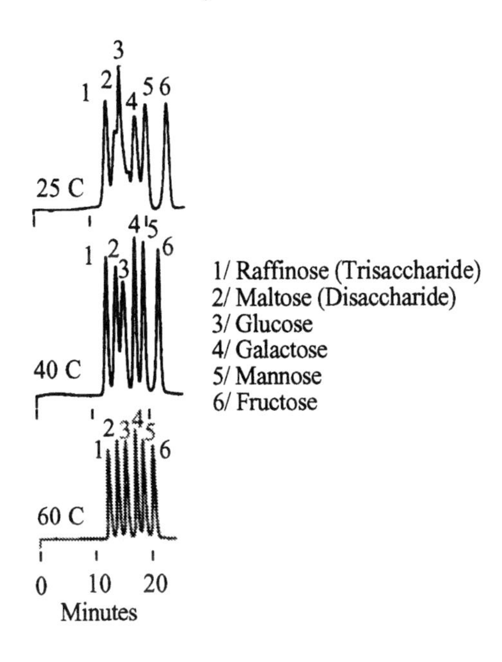

Courtesy of Supelco Inc.

This impressive effect of temperature on efficiency and resolution is not common and improvements of this order of magnitude are not always realized by raising the column temperature. Nevertheless, temperature is a variable that needs to be considered depending on the type of mixture being separated.

Summary

The examples given in this chapter illustrate how the basic mechanisms that control solute retention and, consequently, sample resolution can be changed and varied to achieve a particular separation. The choice of the best phase system for a particular LC analysis is not an art but can be deduced from the physical and chemical characteristics of the solutes that are present in the mixture. The appropriate mechanism and/or the type of molecular interactions that needs to be exploited to complete the analysis follows rationally from the nature and chemistry of the sample components. Experience, without doubt, helps a great deal in phase selection but by applying a logical approach the novice can also arrive at the same solution but perhaps take a little longer.

In general, the majority of separations are achieved by exploiting *dispersive* interactions in the stationary phase and modifying and controlling the absolute and relative retention of the solutes by adjusting the composition of the mobile phase. It is far easier to adjust the mobile phase by selecting different mixtures of water and the solvents methanol, acetonitrile and/or tetrahydrofuran than change from column to column.

Finally, it is worth emphasizing once again that the C8 reverse phase, with a 3 μ particle size, packed in a column 3 cm long and 4.6 mm in diameter is an excellent 'scouting' column. A column of this size can be made to provide very rapid separations and subsequently can be quickly reconditioned to another mobile phase. By using such a column, and employing a gradient from pure water to pure acetonitrile to develop the separation, the complexity of the sample will often be revealed, and from the results an improved phase system can be educed.

Index

acetaminophen/aspirin tablets
  analysis of, 215
  impurities in, 307
acquisition, data, 152
adsorption isotherm
  bi-layer, 63, 64
  chloroform, 60
  ethyl acetate, 64
  mono-layer, 58
  octane sulfonate, 80
  reverse phase, 78
aflatoxins in corn meal, 217
alcohols, aliphatic, reverse phase
    isotherms, 78
alkali and alkaline earth cations
  detection by electrical
    conductivity, 178
  separation of, 178
alkoxy silane reagents, 73
analysis
  computer print-out, 278
  qualitative, 251
  quantitative, 251, 265
    procedures for, 267
    references for, 267
anti-microbial reagents, separation of, 29
applications of LC, 281
aqueous solvent mixtures, 82
area, peak, 12
aromatics
  detection, diode array, 174
  separation of, 27
asymmetry, peak, 111, 252
  from overload, 113
automatic extraction
  apparatus for, 206
  and concentration, 205

base line, 11
  peak width, 12
Beer's Law, 166
benzoyl chloride, 242
1-benzyl-3-p-tolyltriazine, 243
bi-layer adsorption, 63
  ethyl acetate isotherm, 64
blood
  dispersive contaminants, 209
  trimethoprim content, 225
t-boc-alanine, separation of isomers, 295
bonded phase, 71
  elution chromatography, 81
  history, 72
brain, rat, histamine content, 230
brush phase, 73
bulk phase, 75
  reverse, 28

cadmium vapor lamp, 168
capacity ratio, 41
carbon stationary phases, 76
carboxylic acids, UV adsorption reagents, 242
cartridge, solid phase extraction (SPEC), 200
centrifugation, sample preparation, 197
chemical forces and interactions, 23
chiral chromatography, 38, 290
chloroform, adsorption isotherm, 60
chlorosilane reagents, 72
choice of mobile phase, 235
chromatogram
  anti-microbial agents, 29
  of aromatics, 27
  characteristics of, 10
  exclusion, 35
    high efficiency, 37
  reduced, 106
  Warfarin, 39
chromatograph
  contemporary, 123
  liquid, 123
    basic, 124
chromatography
  chiral, 38
  classification of, 7, 8
  exclusion, 34
  history, 2
  introduction to, 1
  nomenclature, 9
  supercritical, 7
citrus fruits, fungicide content, 219
classification of chromatography, 7, 8
column
  definition and function, 15
  diameter, 116
  efficiency, 43
    equation, 45, 46
  optimized, 115
  oven, 144
  overload, peak asymmetry, 113
  preparative, 117
  selection, 114
  switching, 142
  thermal effects, 254
computer print-out of analysis, 278
concentration, sample
  automatic, 205
  preparation, 198
    apparatus for, 206
    sensitivity of, 208
connecting tubes, 149
corn meal, aflatoxin content, 217
corrected retention time, 11
corrected retention volume, 11, 40
critical pair, 107
critical pressure, 7
critical temperature, 7
cross-linked polystyrene, 85
curve, elution, equation, 20
Cyclobonds for chiral separations, 291
cyclodextrin, separation/detection, 187

dansyl chloride, 238
  reagents, 239
data acquisition and processing, 152
dead time, 11
dead volume, 11
de-convolution peak, 273
derivatization, 237
  post-column, 245
  pre-column, 238
  reagents, 238
    fluorescence, 238
    UV adsorption enhancers, 241
detector, 149, 157
  electrical conductivity, 176
  flow sensitivity, 165
  fluorescence, 180
  linearity, 158
    effect of, 160
  multifunctional, 189
    performance of, 191, 192
  noise, 162
    drift, 163
    long term, 163
    measurement of, 163
    short term, 162
  pressure sensitivity, 164
  refractive index, 184
  sensitivity, 164
  specifications, 158
  temperature sensitivity, 165

[detector]
UV, 165
diode array, 171
fixed wavelength, 167
multi-wavelength, 169
diameter, column, 116
diaphragm pumps, 137
diazeprines, separation of, 301
diffusion, longitudinal, 99
equation, 100
3,5-dinitro-benzoyl chloride, 242
3,5-dinitrofluorobenzene, 242
2,4-dinitrophenylhydrazine, 244
diode array UV detector, 171
aromatic detection, 174
peak confirmation, 173
dispersion, 16
tube connections, 149
dispersive interactions, 28
dispersive interaction chromatography, 297
displacement interactions, 61, 65
distribution coefficient
control of, 23
definition, 5
drift, 163
dynamic range, linear, 161

efficiency
effect of temperature, 145
calculation for specific resolution, 109
equation, 110
column, 43
equation, 45, 46
electrical conductivity detector, 176
cell design, 177
elution curve
equation, 20
differential equation, 19
enthalpy of exchange, 30
entropy of exchange, 30
equation
elution curve, 20
maximum sample volume, 96
retention volume, 22
Van Deemter, 97, 103
ethyl acetate adsorption isotherm, 64
evaporation, sample preparation, 199
exclusion chromatogram, 35
exclusion chromatography, 34, 282
with polystyrene gels, 286
with silica gels, 283
exclusion media, silica gel, 67
explosives, separation of, 299
external loop sample valve, 141
external standards, 270
extraction
automatic, 205
apparatus for, 206
sample preparation, 198
sensitivity of, 208

fatty acids in margarine, 213
ferredoxin, separation of, 314
filtration, sample preparation, 196
fixed wavelength UV detector, 167
flavones in grapefruit juice, 229
flow sensitivity, detector, 165
"fluorescamine", 240
fluorescence detector, 180
fluorescence reagents, 238
dansyl chloride, 238
"fluorescamine", 240
"fluoropa", 240
NBD chloride, 239
"fluoropa", 240
forces
chemical, 23
dispersive, 28
ionic, 24
polar, 25
free energy, 31
fructose/glucose in lemon juice, 222
fungicide content
citrus fruits, 219
shampoo, 223

galaction, separation/detection, 188
gas chromatography, introduction of, 4
glucose/fructose in lemon juice, 222
glyceryl-propyl bonded phase for protein separations, 316

gradient programmer
  high pressure, 125
  low pressure, 126
grapefruit juice, flavone content, 229
grinding, silica gel, 57
growth regulators, separation of, 303

heat transfer tube, 148
height, peak, 12
herbicide
  Diquat and Paraquat, separation of, 310
  triazine in pond water, 227
HETP, 104
  curve, 105
  equation, significance of, 105
high pressure gradient programmer, 125
histamine in rat brain, 230
history of chromatography, 2
hormones, steroid, separation of, 308
HPLC, 3
hydrogel, 57
hydrophilic interactions, 53
hydrophobic interactions, 52
hydroxyl groups, silica surface, 58

identification methods, 40
injection point, 10
interactions
  chemical, 23
  dispersive, 28
  displacement, 61, 65
  hydrophilic, 53
  hydrophobic, 52
  ionic, 24
  lyophilic, 53
  lyophobic, 53
  molecular, 23
  polar, 25
  reverse phase, 77
  sorption, 61, 65
interactive LC systems, 296
internal loop sample valve, 140
internal standards, 268
introduction, chromatography, 1
ionic interactions, 24
ionic interaction chromatography, 309
ion pair reagents, 24
ions, separation of, 25
ion suppression, 87
isotherm
  alcohols on reverse phase, 78
  chloroform (mono-layer), 58
  ethyl acetate (bi-layer), 64
  octane sulfonate, 80

jargon, 51
jet mill, 57

lamp, vapor, 168
Langmuir adsorption isotherm, 59
lemon juice, fructose/glucose content, 222
linearity, detector, 158
  dynamic range, 161
  effect on accuracy, 160
line, base, 11
liquid chromatograph, 123
  basic, 124
liquid chromatography (LC)
  applications, 281
  history, 2
liquid chromatography (LC) pumps, 128
liquid-liquid chromatography, 3, 9
liquid-solid chromatography, 9
longitudinal diffusion, 99
  equation, 100
long term noise, detector, 163
low pressure gradient programmer, 126
lyophilic interactions, 53
lyophilization, sample preparation, 199
lyophobic interactions, 53

macro-porous polymers, 84
margarine, fatty acid content, 213
mass overload, 120
mass transfer
  mobile phase, 100
    equation, 101
  stationary phase, 101
    equation, 103
mercury vapor lamp, 168
metanephrine in urine, 232

methadone, separation of isomers, 295
p-methoxy-benzoylchloride, 242
methyl isocyanate, 241
migration of solute, 6
mill, jet, 57
minimum detectable concentration, 164
mixer, post-column reactor, 246
mobile phase
  choice of, 235
  supply system, 124
molecular interactions, 23
mono-layer adsorption, 58
  chloroform, 60
  isotherm, 59
multi-functional detector, 189
  performance of, 191, 192
multi-path dispersion, 98
  equation, 98
multi-wavelength UV detector, 169

NBD chloride, 239
nitro-aromatics, separation of, 27
nomenclature
  chromatogram, 10
  chromatography, 9
noise, detector, 162
  drift, 163
  effect on retention measurement, 264
  long term, 163
  measurement, 163
  short term, 162
non-return valve, 129
normalization, 271
normetanephrine in urine, 232
norphenylephrine, separation of
    isomers, 295

octane sulfonate, adsorption
    isotherm, 80
oligomeric phases, 74
optimized columns, 115
optimum velocity, 105
oven, column, 144
overload, column
  mass, 120
  peak asymmetry, 113
[overload, column]
  volume, 118
    equation, 118

packing, slurry, 58
pair, critical, 107
paired ion reagents, 79
Paraquat, separation from Diquat, 310
particles, irregular, 57
peak
  area, 12
    measurement, 266
  asymmetry, 111, 252
  de-convolution, 273
  height, 12
    measurement, 265
  unresolved, 256
  width, 12
    at base, 12
    at half height, 12
phase
  bonded, 71
  brush, 73
  bulk, 75
  carbon, 76
  oligomeric, 74
  reverse, 76
    interaction, 77
phase ratio, 41
  systems, 51
phenol in river water, 234
phenylephrine, separation of isomers, 292
phenylisocyanate, 241
phthalate esters, separation of, 287
PIC reagents, 79
pigments, 2
plant pigments, 2
Plate Theory, 17
pneumatic pumps, 128
point, dead, 10
point, injection, 10
polar interactions, 25
polar interaction chromatography, 304
polymers
  elution chromatography, 90
  macro-porous, 84
  polystyrene, 85

polysiloxanes, separation of, 68
polystyrene standards, separation of, 284
pond water, triazine content, 227
post-column derivatization, 245
precipitation, sample preparation, 200
precision
of peak area measurement, 272
of peak height measurement, 272
of retention measurement, 263
pre-column derivatization, 238
preparation, sample, 195
preparation techniques, 195
preparative columns, 117
pressure, critical, 7
pressure sensitivity, detector, 164
print-out of analysis, 278
priority pollutants, detection separation, 182
process of migration, 6
processing, acquisition of data, 152
programmer, gradient
high pressure, 125
low pressure, 126
proteins, separation
ion exchange, 311
mixed interactions, 316
non-porous resins, 88, 288
reverse phase resins, 89
protocol, sample preparation, 211
pump
diaphragm, 137
LC, 128
pneumatic, 128
rapid refill, 133
single piston, 132
syringe, 131
twin headed, 135

qualitative analysis, 251, 252
quantitative analysis, 251, 265
procedures for, 267
references for, 267
from retention times, 257

range, linear dynamic, 161
rapid refill pump, 133
rat brain, histamine content, 230
Rate Theory, 94
ratio
capacity, 41
phase, 41
separation, 42
reagents
dansyl, 239
fluorescence, 238
ion pair, 24
UV adsorption enhancers, 241
reciprocating single piston pump, 132
reduced chromatogram, 106
references, 13, 47, 91, 121, 155, 193, 247, 279
refractive index detector, 184
reporting results, 277
resistance to mass transfer
mobile phase, 100
equation, 101
stationary phase, 101
equation, 103
resolution, 108
efficiency required, 109
equation, 110
results, reporting, 277
retention, 15
thermodynamics of, 29
time, 11
corrected, 11, 40
peak composition, 257
volume, 11
equation, 22
effect of solvent composition, 262
effect on temperature, 260
precision of measurement, 263
reverse phase, 76
alcohol adsorption isotherms, 78
bulk, 28
interaction with solvent, 77
river water, phenol content, 234

saccharides
separation/detection, 186
ion exchange, 313
sample
preparation, 195

[sample; preparation]
liquids, 221
liquid-solid, 228
protocol, 211
solids, 212
techniques, 195
taking, method of, 211
valve, 138
external loop, 141
internal loop, 140
volume, equation for maximum, 95
scopolamine, separation of isomers, 292
selection, column, 114
selectivity, 15
sensitivity, detector, 164
flow, 165
pressure, 164
temperature, 165
separation, 93
separation ratio, 42, 293
serum
dispersive contents of, 209
tricyclic antidepressant content, 204
trimethoprim content, 225
shampoo, Suttocide® content, 223
short term noise, detector, 162
silica gel
as exclusion medium, 67, 70
grinding, 57
production of, 55
as a stationary phase, 69
silicic acid, 55
siloxanes, separation of, 68
single piston pump, 132
size, exclusion, 34
slurry packing, 58
solid phase extraction cartridge (SPEC), 200
solids, sample preparation, 212
solute
identification, 40
interaction with silica, 61, 65
migration, 6
solvent composition, effect on retention volume, 262
sorption, 62, 65
SPEC (solid phase extraction cartridge), 200
specifications, detector, 158
standards
external, 270
internal, 268
stationary phase, 33
silica gel, for elution, 69
silica gel, for exclusion, 69
steroid hormones, separation of, 308
summation of variance, 94
supercritical chromatography, 7
supply system, mobile phase, 124
suppression, ion, 87
surface hydroxyl groups, 58
Suttocide® in shampoo, 223
switching, column, 142
syringe pump, 131

tablets, acetaminophen/aspirin, analysis of, 215
temperature
critical, 7
effect on efficiency, 145
effect on resolution, 319
effect on retention volume, 260
sensitivity, detector, 165
tetrahydrocannabinol in urine, 202
Theory, Plate, 17
Theory, Rate, 94
thermodynamics of retention, 29
thermal effects in a column, 254
thiabendazole in citrus fruit, 219
time
dead, 11
retention, 11
corrected, 11
T-mixer, post column reactor, 246
triazine
in pond water, 227
UV chromophore reagent, 243
tricothescenes, separation of, 306
tricyclic antidepressants in serum, 204
Tridet detector, 189
performance of, 191, 192
2,2,2-trifluoro-1-(9-anthryl) ethanol separation of isomers, 317
trimethoprim in blood serum, 225

tube
  connecting, 149
  heat transfer, 147
twin headed pump, 135

unresolved peaks, 256
urine
  metanephrine content, 232
  tetrahydrocannabinol content, 202
UV adsorption enhancers, 241
UV detector, 165
  diode array, 171
  fixed wavelength, 167
  multi-wavelength, 169

valve
  non-return, 129
  sample, 138
    external loop, 141
    internal loop, 140
Van Deemter equation, 97, 103
variance
  per unit length, 104
  summation of, 94
velocity, optimum, 105
vinyl benzene cross-linked polymers, 85
volume
  corrected retention, 40
  dead, 11
  overload, 118
    equation, 118
  retention, 11
    equation, 22
  sample, maximum, 95

warfarin, separation of enantiomers, 39
water
  triazine content, 227
  phenol content, 234
width, peak, 12
  at base, 12
  at half height, 12

xerogel, 57

zinc vapor lamp, 168